JN439866

Weather Forecasting

A Practical Guide for Internet Users

Woo-Jin Lee

KwangGyo E-tax

First published 2011
by KwangGyo E-tax Publishing Company
27-35, Changjeon-dong, Mapo-gu, Seoul, Korea

Weather Forecasting - A Practical Guide for Internet Users

ISBN 978-89-6477-024-5 93440

Printed in the Republic of Korea.

Cover page : First Korean Communication Ocean Meteorological Satellite (COMS) visible image at 02:15 UTC on 12 July 2010. Source: Korea Meteorological Administration.

Weather Forecasting

A Practical Guide for Internet Users

Preface

Our lives are sensitive to changes in the weather and climate. Dozens of episodes easily come to mind when outdoor activities were hampered by sudden forces of Mother Nature. Picnics were cancelled by dust storm warnings. Baseball games were halted by heavy rains. Schools were closed due to heavy blizzards. Roads became impassable due to flash floods. Holiday tickets were rendered useless due to an approaching tropical storm at the destination.

Our society becomes more and more vulnerable to the extremes in weather and climate as the population grows and urbanization extends. The environment is degraded with the accumulation of man-made pollutants, desertification, deposition of acid rain, and ozone layer depletion. Greenhouse warming and associated climate change may further threaten our lives and properties due to stronger and more frequent storms and weather and climate extremes.

For a long time the business sector has recognized the importance of weather and climate information as a strategic resource to mitigate potential risks and to improve efficiency and productivity in various economic sectors including agriculture, forestry, water supply and flood management, transportation, energy, public health, disaster insurance, and emergency management. Some industries such as construction, tourism, and sports are more concerned in designing alternatives to respond to foul weather events; other businesses such as electricity, daily produce and agro-livestock harness weather information to improve productivity and to reduce cost of preventive measures. Some sectors, such as media, sell the weather as a story. Some industries such as the future markets, international trading, and even reinsurance look up the uncertainty of weather and climate predictions.

One can access weather information from various sources, such as television, radio, newspaper, mobile phones, and the Internet. Particularly, the Internet has many advantages compared with other media sources. First, it is interactive. You can seek the information you want at any time and at any place on various subjects that range from short and long-term forecasts, observations, weather terminology, to learning materials. Second, information is updated in real time. You can get the latest information that occurs in the atmosphere around the world from radar and satellite animations to severe weather warnings. Third, the Internet provides rich references for further study. You can find scientific papers, articles, reports on the specific subjects of the weather process and the mechanisms working behind it. Finally, it is multi-perspective. You can hear multiple voices on any single subject. If you have a proper knowledge and sound judgment, you could have a better stance to review various options for decision.

Above all, the most fascinating point of Internet is that it is free of charge in most cases. The author has seen many people who want more accurate weather forecasts but could not afford to spend money to buy them. Most of them access the weather information from television, radio, and newspaper. However, such public information is never sufficient to answer the details of weather evolution in terms of locality and timing. The author himself has been using the Internet most of time to get the current short and long-term weather forecasts, and to learn the latest knowledge and experiences on the rapidly developing science of weather and climate. It is noted that most free sites found on the Internet hardly provide customized information for your own personal problems concerning the weather. In this book, the author would like to address the practical guidelines, based on his own experience, on how to get proper information through the Internet, how to interpret them, and how to apply them during decision-making situations.

The science of weather forecasting advances very rapidly thanks to the supercomputing infrastructure, numerical weather prediction models, and data assimilation of space borne observations. However, the computer forecasts on the Internet have certain degrees of errors, as they are produced from the imperfect initial conditions and incomplete models. The forecasters spend more and more time on the examination of computer generated outputs for the daily weather forecasting, and the understanding of computer forecast models and their outputs become an indispensible part of the forecaster's training. Many parts of this book are devoted to explain the characteristic of computer model outputs with numerous case examples. Special focus is given on the use of computer model guidance on the high impact weather in the time scale of a week, and extended to long-range climates whenever a common approach can be applicable for both.

Chapters 1 and 2 contain background knowledge on meteorological information available on the Internet, including weather warnings and forecasts. Chapters 3 and 4 discuss the interpretation of basic forecast charts for daily weather and seasonal climate. Chapters 5 and 6 elaborate on the limitation and uncertainty of forecasts, and the application of probabilistic forecasts based on ensemble prediction systems. Chapter 7 explains the background knowledge and interpretation of long-range forecast charts. Chapter 8 discusses the characteristics of track and intensity forecasts of extratropical and tropical cyclones. Chapters 9 touches on the mesoscale model products and their limitations. Finally, Chapters from 10 to 13 deal with the correction and refinement of model forecasts to derive sensible weather elements.

In touching upon the diverse subjects from the warning symbols to the model forecast charts, the scope and depth of the contents are not uniform throughout the various chapters. This book is designed for the general readers who have a minimal background in meteorology but are interested in using meteorological information available on the Internet. However, part of the chapters from 9 to 13 requires some knowledge on general meteorology and numerical weather prediction. The beginners may skip those chapters without losing logical ground.

The author uses weather situation mostly from Asia as examples in this book, simply due to the familiarity and convenience the author has in using them. The figures in the book are excerpts from the Internet, even though some websites can only be accessed with a password. As web content constantly evolves and changes, some of the URLs that appear in the book may no longer valid by the time of publication. Whatever URL changes or content updates, the essence of the guidelines in this book hopefully remains the same.

Acknowledgements

I would like to express special thanks to Prof. Kenneth C. Crawford and Dr. Ji-Yung Kim for the valuable comments and suggestions on the manuscript. My warm thanks are due to Dr. Bill Bourke and Ms. Mara Baviera for reviewing an earlier version of the manuscript and for providing informative suggestions for improvement. I appreciate that Dr. ChunHo Cho, Dr. Hoon Park and colleagues at Numerical Prediction Office helped me proofread the manuscript. I would like to thank to Mr. Young-Hwa Kim for the thermodynamic diagrams, and Mr. Yong-Su Kim for the figures 7.2.1, 7.2.2 and 13.1.3. I am grateful to Ms. Su-Kyeong Park for the editorial advice. The encouragement and support of Dr. Byoung-Sook Ko, President of Kwanggyo Etax Publishing Company, is gratefully acknowledged.

The Unified Model charts introduced in this book are produced from the CRAY supercomputer at the Korea Meteorological Administration. I am greatly indebted to KMA for the opportunity to access the numerous figures, part of which appear in this book. Finally, I would like to thank to Dr. Mark Hobson at CIMSS, Dr. Manfred Kloeppel at ECMWF, an anonymous correspondent at EPA, Dr. Naoyuki Hasegawa at JMA, Dr. Elizabeth Lessard and Dr. Greg Thompson at NCAR, Dr. Jerry Slaff at NOAA, and Dr. Judith C. C. Torres and Dr. Koji Kuroiwa at WMO for their advice on the use of the graphic material presented on their websites.

Abbreviations

4dVar	4-dimensional Variational data assimilation.
AAO	AntArctic Oscillation.
AC	Anomaly Correlation.
ADAM2	Asian Dust Aerosol Model version 2.
AO	Arctic Oscillation.
APCC	APEC Climate Center.
ARL	U.S. NOAA Air Research Laboratory.
ASC	Aircraft-Satellite Comparisons.
ASCAT	European Advanced Scatterometer.
AVHRR	Advanced Very High Resolution Radiometer.
AWS	Automatic Weather Station.
BCC	Beijing Climate Center.
BMJ	Betts-Miller-Janjic convection scheme.
BoM	Bureau of Meteorology.
BSS	Brier Skill Score.
CAPE	Convective Available Potential Energy.
CAPPI	Constant Altitude Plan Position Indicator.
CAS	WMO Commission for Atmospheric Sciences.
CBS	WMO Commission for Basic Systems.
CIMSS	Cooperative Institute for Meteorological Satellite Studies in University of Wisconsin–Madison.
CIN	Convective Inhibition.
CIRA	Cooperative Institute for Research in the Atmosphere.
CMC	Canadian Meteorological Centre.
COMET	Cooperative Program for Operational Meteorology, Education and Training.
COMS	Korean Communication Ocean Meteorological Satellite.
CP	Convective Parameterization.
CPC	U.S. Climate Prediction Center.
CPTEC	Centro de Previsão de Tempo e Estudos Climáticos.
CSI	Critical Success Index.
DD	Dew-Point Depression.
DEMETER	Development of a European Multimodel Ensemble system project.

ECMWF	European Center for Medium-range Weather Forecasts.
EFI	Extreme Forecast Index.
ENSO	El Niño Southern Oscillation.
EPA	US Environmental Protection Agency.
EPS	Ensemble Prediction System.
ESA	European Space Agency.
EUMETNET	Network of European Meteorological Service.
FAR	False Alarm Ratio.
FY	Feng Yun Chinese satellite.
GCM	General Circulation Model.
GDPFS	Global Data-Processing and Forecasting System.
GIS	Geographic Information System.
GOES	Geostationary Operational Environmental Satellite.
HMC	U.S. NCEP Hydrometeorological Center.
HYSPLIT	HYbrid Single-Particle Lagrangian Integrated Trajectory model.
IBTrACs	International Best Track Archive for Climate Stewardship.
INMET	Instituto Nacional de Meteorologia.
IPCC	International Panel on Climate Change.
IPV	Isentropic Potential Vorticity.
IR	InfraRed channel imagery.
IRI	International Research Institute for global change studies.
ISO	International Organization for Standardization.
ITCZ	Intertropical Convergence Zone.
JMA	Japan Meteorological Agency.
JTWC	Joint Typhoon Warning Center.
KF	Kain Fritsch convection scheme.
KMA	Korea Meteorological Administration.
Meteosat	European Meteorological Satellite.
MJO	Madden Julian Oscillation.
MM5	Mesoscale Model version 5.
MODIS	American Moderate Resolution Imaging Spectroradiometer.
MOS	Model Output Statistics.
MP	MicroPhysics.
MSLP	Mean Sea Level Pressure.
MSSS	Mean Squared Skill Score.
MTSAT-1R	Japanese Multi-functional Transport Satellite.

NAO	North Atlantic Oscillation.
NCAR	U.S. National Center for Atmospheric Research
NCEP	U.S. NOAA National Centers for Environmental Prediction.
NESIS	Northeast Snowfall Impact Scale.
NETWC	Near-Equator Trade Wind Convergence.
NOAA	National Oceanic and Atmospheric Administration.
NOGAPS	U.S. Navy Operational Global Atmospheric Prediction System.
NWS	U.S. National Weather Service.
PBL	Planetary Boundary Layer.
PNA	Pacific-North American Pattern.
POD	Probability of Detection.
PPI	Plan Position Indicator.
RMSE	Root Mean Square Error.
RSMC	Regional Specialized Meteorological Centers.
SAR	Space-borne synthetic Aperture Radar.
SAWS	South African Weather Service.
SCAN	System for Convective Analysis and Nowcasting.
SPCZ	South Pacific Convergence Zone.
SPF	Sun Protection Factor.
SSC	Satellite-Satellite Comparisons.
SWIC	Severe Weather Information Center.
TCC	Tokyo Climate Center.
TCP	Tropical Cyclone Program.
THORPEX	The Observing System Research and Predictability Experiment.
TIGGE	THORPEX Interactive Grand Global Ensemble.
TUTT	Tropical Upper Tropospheric Trough.
UCAR	University Cooperation for Atmospheric Research.
UKMet	United Kingdom Meteorological Office.
UM	U.K. Met Office Unified Model.
VIS	VISible channel imagery.
WCRP	World Climate Research Program.
WGNE	Working Group on Numerical Experimentation.
WMO	World Meteorological Organization.
WRF	Weather and Research Forecast Model.
WV	Water Vapor channel imagery.
WW3	Wave Watch 3.
WWIS	World Weather Information Service.

Contents

Part I. General Information

Part II. Forecast, Verification and Probability

Part I. General Information

1. Weather on Internet

2. Severe Weather Warnings and Forecasts

1. Weather on Internet

1.1 Global weather at home

One early afternoon in Atlanta, USA, the sun was shining strongly through the autumn sky. Jenny, a trade consultant is packing her clothes. She is bound for Moscow, where she will be heading the next day. She will be spending the next few days on business in the Russian capital. She takes a small bag from the floor and brushes off the dust from the surface and proceeds to list the things she will be bringing with her. After a while, she stops to ask herself if the bag would be large enough to hold the thickness and volume of the clothes she intended to bring. Should I bring an overcoat? Will lighter clothes and a vest be more suitable? How much colder would Moscow be, being so much farther to the north compared to Atlanta? Would it be snowing or raining when I get there? Should I pack an umbrella?

One click on the icon labeled "world weather" at the National Weather Service website and the weather report for the Moscow airport appears on her computer screen. Cloudy and rainy! Five degrees Celsius. Moscow time zone is six hours earlier than Atlanta so the local weather information must be observed at the evening. What will be the maximum temperature tomorrow at Moscow? She continues to search for the weather forecast at Moscow. She is very fortunate, as it is often difficult to find weather forecasts for small cities.

At the same time in Toulouse, a southern city of France, Tom is in a hurry to go home to Seoul, Korea. He is heading to the airport where his flight would take him for a stopover at Paris. He is waiting at the hotel lobby for the rain to stop. A message has just been broadcasted from the Toulouse airport saying that all flights are delayed due to severe thunderstorms nearby. Immigration and baggage control will be difficult to get through, the Paris airport being a hub for international flights. There is every reason that the boarding process would be delayed, if not for the weather. Not being on the scheduled flight to Paris means he would have to stay at the French capital for a day or two to find out schedules for the next connecting flight. If he could not take off at all from Toulouse, then he has to stay at Toulouse for another night and to extend the arrangements for accommodations accordingly. Raindrops fall against the window and the occasional thunder roars from the dark sky as he nervously waits. He decides to open his computer and search the portal sites for the weather at Toulouse. Météo-France appears on the screen, but everything is written in French. Weather flashes in front of him. Satellite images and radar loops pop up. If only he knew how to interpret the images, he could tell if the rainstorms were moving in or out of the airport. He scratches his head, and searches other websites. Local radar images could not be found anywhere, as it is difficult to find very short-range forecasts in real-time on the Internet. Tom would have to drive for the airport leaving his destiny to fortune.

Cheol-Soo, living at Daejeon, Korea, is excited for a three-day trip to Baekryeong-do Island, which is located 100 km off the west coast of the border between North and South Korea. He has to decide on the exact dates for the trip next week. And yet he is worried about the work piling up at the office. The cruise to the island often halts for several days on rough seas. Travelers would have to wait at the port until the winds attenuate. Sea sickness is another concern in the tumbling waves. Fortunately, the forecast chart for mean sea level pressure is available for the next week along with the local wave forecasts. If he were able to interpret the surface charts to forecast the wave height, he himself could wisely choose when to go for the trip and avoid rough seas.

Peter is busy organizing a football match for the homecoming day at Hessel Park, Urbana-Champaign. The match is just a week ahead; and he is a bit depressed by the weather forecast saying there will be rain on that day. Two scenarios come to Peter's mind : Cancelling the match would mean he has to give prior notice to the participants well ahead of time. Carrying on with the plan means he has to pay the fees by today. Light rain may not significantly influence the game. Moderate amounts may be okay if it comes early in the day and stops before noon. Looking for the probability of precipitation forecast from the local meteorological center website, he discovered that there would be 40% chance of raining on that day! What should he do? How reliable is the forecast? If he could interpret the weather charts and compare forecast charts from several meteorological centers, this would increase his confidence in predicting the likelihood of rain on that particular day.

Soon-Hee is going to get married in a few months' time. She wants to choose a sunny day at Jeju Island in Korea. She sits in front of her computer, and enters the KMA website for long-range weather forecasts. The seasonal forecast says that rainfall amount would be above normal and the mean temperature would be below normal. She could not find a daily forecast beyond a week ahead. She accesses the climatological database. She would be in a better position, if she chooses a date when fair weather conditions have been consistently observed in the past for that day of a year.

The above cases are only a few examples from millions of daily life occurrences in any part of the world: a salesman choosing business suit for brunch, parents handing over umbrellas to kids on the way to school, a coordinator planning a golf tour or mountain climbing for the weekend, a manager supervising outdoor events, and a technical team leader checking the status of tires of racing cars. They may laugh or cry depending on the weather forecast and its accuracy.

Most people access meteorological information through television, newspaper, and telephone calls. Information obtained through such means is so general that it could not meet the detailed needs and requirements of each individual. On the other hand, the Internet is becoming more popular than ever. More information can be found through the Internet than what we expect. Some of these websites are very simple and easy to understand. Others are very sophisticated, requiring professional training to digest and convert the information into practical knowledge.

The World Meteorological Organization (WMO) supports the open information policy, and encourages the exchange of meteorological information among member countries as long as it contributes on the protection of life and reduction of property damage against severe weather and climate conditions. However, each country applies different policies on the distribution of information through the Internet. Open information policy is strongly supported in the United States. Private companies could access most of the weather information produced by U.S. National Weather Service (NWS) with minimum costs covering communication fees, which promotes the commercialization of the meteorological service. A similar policy can be found in Korea. On the other hand, many other countries including most of Europe protect meteorological information from free access and instead encourage commercial sectors to sell the weather information in the form of value-added products. Some countries have positions in between those two extremes. For instance, a non-profit data center collects the information from the Japan Meteorological Agency (JMA), which in turn provides the service to the private sector. The Bureau of Meteorology (BoM), Australia used to have a consulting department that produces value added products with consulting fees.

Whatever policy is applied for the delivery of meteorological information, the basic meteorological information can be available and downloaded from the Internet without charge. Travelers in United States could get as much information from the U.S. NWS website as they need. Others traveling over Australia could also find plenty of information on the website of BoM. The Internet portal sites provide weather information in real time, and links to local or remote weather sources. Many private websites provide their own weather service products or links to the local weather conditions and short-term forecasts. The major broadcasting companies and newspapers do the same thing. It is not surprising to learn that many research institutes, universities, and private companies also provide their own weather information and links to national meteorological centers. The value of the information depends on the user's ability to exploit it. If the users are able to choose the appropriate source of information and interpret the weather charts and relevant symbols, they could forecast the weather in advance at any part of the world.

As information technology advances and the market pressure to reduce costs intensifies, the number of the users increases who would like to apply the weather and climate information for decision-making in business situations. A one-degree temperature increase results in huge sales profits of beer and ice cream. A correct forecast of the maximum temperature could put one well ahead of competition by securing more stocks well before demand peaks. One could protect vegetation from frost by forecasting the minimum temperature for the following morning. Anticipating rainy days in the next few days would allow for advanced sales and harvest, thereby reducing expected losses from the spoiled crops. Foreseeing rains for the following day at the field office, would enable one to withdraw a contract for painting apartment walls and save on manpower expenses and other resources. A person who could read the weather map and satellite images may stand in a better position to go about business and daily life compared to others who could not.

It is too time-consuming to analyze all the information available and to evaluate their credibility. Big enterprises hire professionals to consult on the weather. However, small companies have to rely on their own expertise in interpreting weather information for their business. This book is written to illustrate what data, charts, and information to search for, how to evaluate, and how to interpret meteorological information accessible through the Internet. Plain language is used to address the main points without losing some scientific background. Little knowledge is required to understand these primary concepts. The author will be grateful if this book stimulates interest on the interpretation of meteorological information from the Internet and develops expertise on the application of this information during daily life and business.

1.2 Satellite and radar imagery

Among the weather elements, rain and snow have the most impact on the daily life. It is of interest not only when and where precipitation would occur, but also how much. Clouds are composed of small water droplets, which eventually fall in the form of rain or snow when appropriate physical conditions are met. Our sense of cloudiness depends on the height of cloud base as well as the thickness of clouds. One feels fair under the overcast sky if the sunshine partially passes through the high-level clouds. On the other hand, one feels the dark foreboding of the sky when it is obscured by low-level clouds. Most heavy rains fall from thick clouds with a low-level cloud base. Only light rain falls from high-level clouds.

The World Weather Information Service (WWIS) maintained by WMO provides the global view of clouds, rain and snow updated every 3~6 hours with a time lag of 40~90 minutes for data processing. Clicking on the menu bar on the left hand side of the main window of WWIS to select the cloudiness and rain/snow, the global report appears on the screen. One could zoom-in to inspect weather reports for a specific region in the map. More advanced users may continue to read the rest of this section describing how to read weather symbols; otherwise the readers may skip to the next section.

The cloud pictures from meteorological satellites typically are the first data to inspect. This data tells us where the clouds are, what approximate height levels they occupy, what direction they are moving to, and how intense they are. Satellites view the clouds from above through different channels of radiation. Clouds are very sensitive to changes in atmospheric state, and the interpretation of satellite imagery is the first step in grasping what's going on in the atmosphere. Traditionally clouds are represented by bright colors against a black background. Commonly used channels are from the visible and infrared range in the light spectrum. Many meteorological centers provide various satellite imagery on the web, including visible (VIS), infrared (IR), and water vapor (WV) channel imagery.

Visible imagery

During daytime, the clouds are discernible by bright colors in the visible image because they reflect the incoming solar radiation. Thicker clouds reflect more sunlight, and are shown as brighter than the thin clouds. Cloud textures and patterns are easily identified. Cumuliform clouds and stratiform clouds appear in the form of cellular pattern and flat structure respectively. However, it is hard to distinguish the low clouds and snow from high clouds in the VIS image, as they have similar reflectivities. The left panel of Fig. 1.2.1 illustrates an example of visible image from MTSAT-1R geostationary satellite. A series of cloud clusters are developed from the Bay of Bengal to the Philippine Sea along the Intertropical Convergence Zone (ITCZ). In mid-latitudes, a remnant of the typhoon Sinlaku passes the central part of Japan, traveling to the East. Another band of clouds stretches over the central part of China.

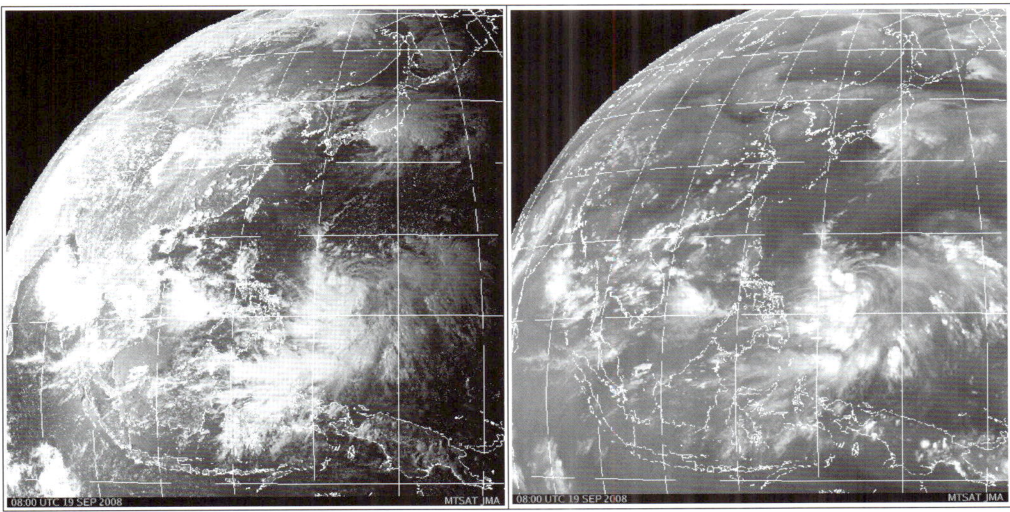

Fig. 1.2.1. Examples of Visible (left) and Water Vapor (right) images at 08 UTC on 19 September 2008 from MTSAT-1R geostationary satellite (http://www.jma.go.jp/en/gms/index.html?area=1&element=2&mode=UTC). Source: Japan Meteorological Agency (JMA).

Infrared imagery

After sunset, IR images are used instead. They represent the temperature field that is equivalent to the radiant energy emitted from the clouds. High clouds, emitting thermal radiation at low temperatures, are recognized by the bright white color in the IR image. The low clouds under the high clouds can hardly be seen in the infrared image. The left panel of Fig. 1.2.2 illustrates an example of infrared image from American Geostationary Operational Environmental Satellite (GOES). The major cloud clusters over Texas are associated with the Hurricane Ike, which made landfall near Houston after crossing the Gulf of Mexico. The bright area near its center indicates well-developed, deep convective clouds; the outward spirals represent the cloud shield at upper levels. Three other groups of clouds are located in the mid-latitude zone. The 2nd group of clouds hangs over Illinois, Indiana and Ohio. The 3rd and 4th groups are found over North Dakota and Montana respectively.

Often, composite images combining both VIS and IR images are used during the daytime to identify low clouds from the high clouds above. One should note, however, that the fog and snow at times look like stationary clouds in the satellite image.

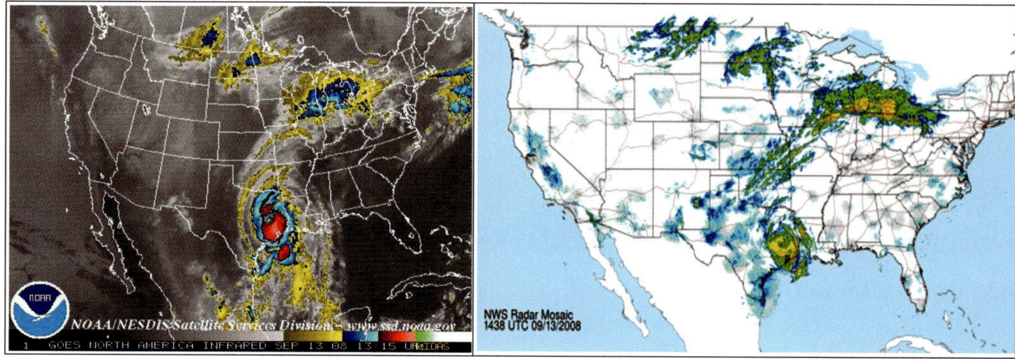

Fig. 1.2.2. Example of (left) Infrared image at 13:15 UTC on 13 September 2008 from GOES

geostationary satellite (http://www.weather.gov/sat_tab.php?image=ir), and (right) corresponding radar reflectivity map at 14:38 UTC on the same day (http://www.nws.noaa.gov/radar_tab.php). The rain bands in the radar reflectivity map reveal a one to one correspondence with the cloud clusters in the IR image. Source: U.S. National Oceanic and Atmospheric Administration (NOAA).

Water vapor imagery

The WV image is also available 24 hours a day as is the IR image. It is particularly helpful in diagnosing the regions of dry and moist air in the middle and upper troposphere. On the grey scale, dark colors are used for lower sensor values indicating dryness in the upper troposphere, often associated with subsidence. White color implies moistness in the upper troposphere. WV images reflect the large-scale flow patterns very well including jet streams, ridges and troughs. The moisture field in the WV image indicates a wave pattern from Northern China to Northern Japan, and appears to link the cloud systems in the VIS image in Fig. 1.2.1. The dark edge of the WV corresponds to the leading edge of the upper level trough, which often signifies the approach of sensible weather. For example, the water vapor image over the central part of Japan has a dark edge intruding from the west. It supports the extratropical transition of the typhoon Sinlaku and the unsettled weather downstream. This point will be revisited in Chapters 4 and 11.

Microwave imagery

Some national centers upload microwave channel images on their website in addition to the basic images. This additional information helps in diagnosing the aerial distribution and intensity of the precipitation systems embedded in the clouds, particularly the eye wall structure of tropical cyclones. Microwave channel images can be accessed from the websites of the U.S. Navy, the Cooperative

Institute for Research in the Atmosphere (CIRA) of the Colorado State University and numerous other sites not mentioned here for brevity.

There are two types of satellites worldwide: geostationary and polar orbiting. GOES, Japanese Multi-functional Transport Satellite (MTSAT-1R), Feng Yun Chinese satellite (FY), European Meteorological Satellite (Meteosat), and Korean Communication Ocean Meteorological Satellite (COMS) belong to the former, while American Moderate Resolution Imaging Spectroradiometer (MODIS) and the European Advanced Scatterometer (ASCAT) are classified as belonging to the latter. The cloud imagery from geostationary satellites is fixed in coverage and updated every 30 or 60 minutes. Polar orbiters take pictures while they are moving along the horizon. It takes several hours for polar orbiters to return to the same position. On the other hand, geostationary satellites are located thousands of kilometers higher above ground compared to polar orbiters. The spatial resolution of the latter is about 1 km, which is twice as high as that of the former.

A single satellite can not provide surveillance across the whole globe. It is always effective to click in the local national meteorological centers' homepage covering the target region of interest. For instance, cloud images over Asia can be easily obtained from any meteorological center website that link the satellite images of FY or MTSAT-1R. The list of contacting address for the national meteorological centers around the world can be found in the WMO website. It should be noted that each national meteorological center uses a different color enhancement scheme to locate the desired energy levels or cloud tops, and to improve interpretation for specific features. More sophisticated and detailed instructions for the interpretation of satellite images can be found at homepages of Cooperative Institute for Meteorological Satellite Studies in University of Wisconsin - Madison (CIMSS), European Space Agency (ESA), JMA, U.S. National Oceanic and Atmospheric Administration (NOAA), U.K. Meteorological Office (UKMet), WMO, and others.

Radar reflectivity

Radar reflectivities provide detailed distributions of on-going rainfall and its intensity. To view the real time radar image at a given locality, the easiest way is to go to any of the local websites that the national meteorological centers maintain. For instance, one may visit the website of Météo-France when traveling to Paris.

Radar is designed to detect the pulses of microwave radiation reflected from raindrops or snowflakes. The radar views the rainstorms in the radial direction with the range of 200~400 km outward. The two-dimensional image from the Internet is produced through the scanning of the atmosphere either by Plan Position Indicator (PPI) or Constant Altitude Plan Position Indicator (CAPPI).

It should be noted that the data from different distances to the radar are at different heights above ground in PPI. The echo patterns in PPI vary with the elevation angle of the beam transmitted from the antenna. CAPPI is basically a horizontal cross-section through radar data that combine multiple PPI scans with different elevation angles. It enables one to view the radar echo on an equal stance at difference distances from the radar. It is necessary to check beforehand if the radar image is produced in either PPI or CAPPI modes, as both have different characteristics. Normally, radar echoes as they appear in most websites are measured at low elevation angles in PPI mode or at low altitudes in CAPPI mode to deliver the information near the ground.

Active convection brings heavy rain or snow, and is identified with the strongest reflectivity and with tight gradients of reflectivity. The stratiform clouds are associated with light rain, and have lower reflectivities. The reflectivity in the radar image appears in certain patterns, such as line-shaped, isolated cells, and aggregated cluster. For instance, both rainbands in North Dakota and Montana mostly come from stratiform type clouds, while those in Texas and the mid-west plains states are partially due to the convective type of clouds in Fig. 1.2.2. The line-shaped rainbands in North Korea, as shown in Fig. 1.2.3, are associated with violent thunderstorms embedded along a cold front.

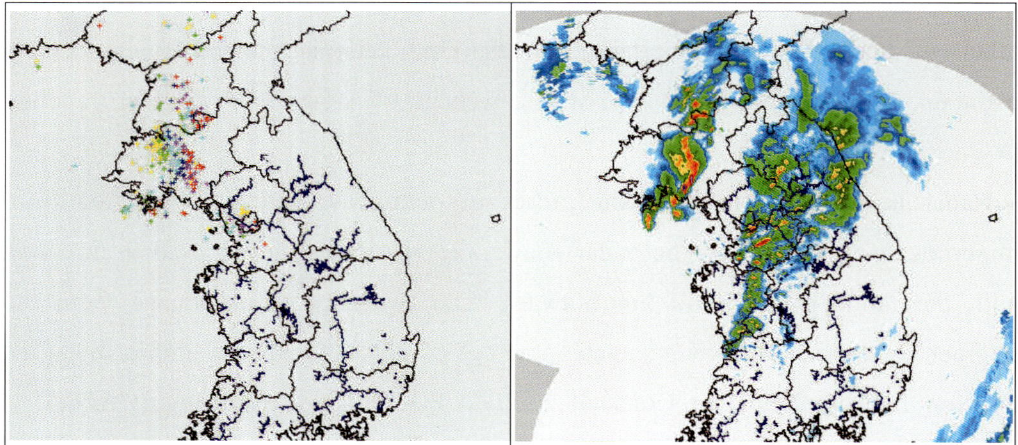

Fig. 1.2.3. Examples of (left) lightning strikes at 18 LST on 12 May 2007 available from http://web.kma.go.kr/eng/weather/images/lightning.jsp, and (right) corresponding radar reflectivity in PPI mode available from http://web.kma.go.kr/eng/weather/images/radar.jsp. False colors indicate time of detection for the left panel, and converted rainfall rate for the right panel. The color bars are not presented for brevity. This image is available from the website of the Korea Meteorological Administration. The collocation of lightning strike points and high reflectivity regions indicate the occurrence of active thunderstorms. Source: Korea Meteorological Administration (KMA).

Z-R relationship

Reflectivity is converted into the precipitation rate through the reflectivity-precipitation transformation function, or the so-called Z-R relationship. Generally speaking, higher values of reflectivity indicate heavier precipitation. Reflectivity in the range of 30~40 dBZ indicates showers or heavy rain. Strong thunderstorms are likely for reflectivity values that exceed 50 dBZ. The relationship, however, is not universal. It varies from cloud to cloud and from season to season, which sometimes causes large errors in the estimation of precipitation rate from the radar returns. Furthermore, the echoes can be often contaminated by noise from various sources, including anomalous propagation, topographic effects, terrain and beam blockage, bight bands, and attenuation. Specifically, when the water-ice mixture is below the freezing level, the radar detects a high intensity return than would otherwise occur for the precipitating clouds. The comparison of radar images with ground observations would help confirm the rain or snow band from the radar echoes. Precipitation rate, typically in units of millimeters per hour, is divided into several classes. Each class has a one-to-one correspondence with a unique color in the radar imagery.

Movie loop

From the animation or movie loops of radar images, one can easily evaluate the speed and direction where the reflectivity patterns are moving. The individual convective cells may move in a different direction from that of the entire system in which they are imbedded. The total rainfall in a region depends not only on the strength of the radar returns but also on the duration of the presence of precipitating clouds affecting the region. The stratiform clouds with weak reflectivity may produce significant amounts of rainfall, if their coverage is large enough to influence a given target region for a longer period of time. Weather specialists use additional radar loops including Doppler radial winds and vertical cross section of reflectivity. However, those miscellaneous products rarely appear in the public websites.

Some national meteorological centers such as Korea Meteorological Administration (KMA) provide lightning strike maps on the web in real-time, as illustrated in Fig. 1.2.3. The area of lightning strikes often is collocated with high reflectivity regions that provide additional evidence for the identification of a severe thunderstorm. It is noted that the radar images in the websites passed through complex quality control processes that suppress various noises associated with the remote detection of cloud and water droplets. More details on the interpretation of radar images can be found at homepages of BoM, NOAA, and numerous others.

1.3 Ground measurements

The state of the atmosphere near the surface of the earth is determined by pressure, wind, temperature, and humidity. Wind speed and direction are measured at 10 m from the ground, and typically averaged over 10 minutes. The other elements are measured at a height of 1.5 to 2 m above ground. The current values of these variables can be accessed at websites of national meteorological centers near one's residence, which are typically updated hourly or every three hours. Some national meteorological centers further upload the observations measured from densely covered automatic weather station (AWS) networks with spatial resolution of roughly 20 km or less. An example is shown in Fig. 1.3.1 for the local map of wind observations as they appear on the JMA website. Similar maps can be found at KMA website as well. Wind observations are useful in diagnosing wind structures of mesoscale systems, and to fix the position of tropical cyclones as they make landfall.

These maps include other surface observation fields such as temperature, humidity and amount of precipitation. They are updated every few minutes. One can even access the vertical profile of winds every minute, based on a limited number of wind profiler observation networks deployed worldwide. Some centers use Fahrenheit and knot, instead of Celsius and m/s for the units of temperature and wind speed respectively.

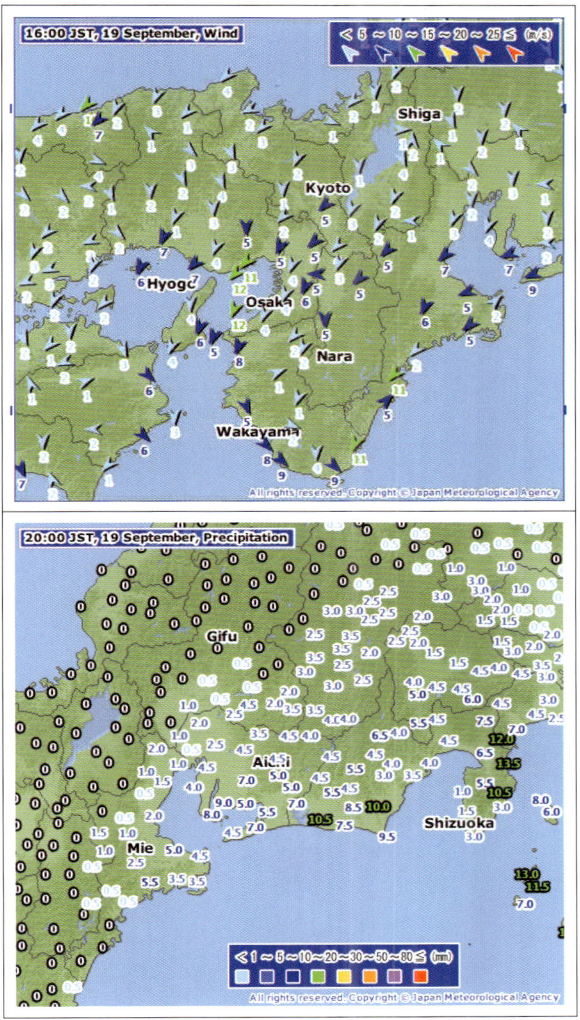

Fig. 1.3.1. (Upper) Surface wind in units of m/s at 07 UTC and (Lower) hourly rainfall in units of mm at 11 UTC on 19 September 2008, as the weakened typhoon Sinlaku traveled eastward near the automatic weather station network of JMA. The wind direction is indicated by arrows. The wind speeds and rainfall amounts are represented by different colors. The image is available from http://www.jma.go.jp/jp/amedas/214.html. Other weather elements such as temperature and relative humidity can be found as well on these sites. Source: JMA.

The wind power is proportional to the square of wind speed. Thus, the strength of wind abruptly increases as the wind speed increases. Wind constantly fluctuates in a chaotic manner, as the air motion is very turbulent. Averaging over 2~10 minutes are used to filter out the noise. It is noted, however, that the instantaneous wind speed, or gust wind, is often more relevant for the assessing damage caused by the wind. The gust wind speed is about 1.66 and 1.23 times higher than the ten-minute average wind speed inland and at sea respectively (Harper et al., 2008).

Wind scale

The Beaufort scale in Table 1.3.1 is designed to measure the wind strength based on familiar observations that can be found in daily life. It is difficult to drive at highway speeds when the wind strength exceeds Beaufort scale 8. When wind strengths exceed Beaufort scale 10, driving becomes dangerous as debris from roofs or wall blocks begins to fly aloft in the wind (http://www.jma.go.jp/jma/kishou/know/yougo_hp/kazehyo.html).

Table 1.3.1. Beaufort scale based on 10 minute averaged wind speeds. This table is reproduced from the WMO manual (http://www.wmo.int/pages/prog/www/WMOCodes/VolumeI1.html#VolumeI1)

Force	Wind km/hr (knots)	WMO Classification	Appearance of Wind Effects	
			On Water	On Land
0	0 (Less than 1)	Calm	Sea surface smooth and mirror-like	Calm, smoke rises vertically
1	1~5 (1~3)	Light Air	Scaly ripples, no foam crests	Smoke drift indicates wind direction, still wind vanes
2	6~11 (4~6)	Light Breeze	Small wavelets, crests glassy, no breaking	Wind felt on face, leaves rustle, vanes begin to move
3	12~19 (7~10)	Gentle Breeze	Large wavelets, crests begin to break, scattered whitecaps	Leaves and small twigs constantly moving, light flags extended
4	20~28 (11~16)	Moderate Breeze	Small waves 1~1.5m. becoming longer, numerous whitecaps	Dust, leaves, and loose paper lifted, small tree branches move
5	29~38 (17~21)	Fresh Breeze	Moderate waves 2~2.5m taking longer form, many whitecaps, some spray	Small trees in leaf begin to sway

Force	Wind km/hr (knots)	WMO Classification	Appearance of Wind Effects	
			On Water	On Land
6	39~49 (22~27)	Strong Breeze	Larger waves 3~4m, whitecaps common, more spray	Larger tree branches moving, whistling in wires
7	50~61 (28~33)	Near Gale	Sea heaps up, waves 4~5.5m, white foam streaks off breakers	Whole trees moving, resistance felt walking against wind
8	62~74 (34~40)	Gale	Moderately high (5.5~7.5m) waves of greater length, edges of crests begin to break into spin drift, foam blown in streaks	Whole trees in motion, resistance felt walking against wind
9	75~88 (41~47)	Strong Gale	High waves (7~10m), sea begins to roll, dense streaks of foam, spray may reduce visibility	Slight structural damage occurs, slate blows off roofs
10	89~102 (48~55)	Storm	Very high waves (9~12.5m) with overhanging crests, sea white with densely blown foam, heavy rolling, lowered visibility	Seldom experienced on land, trees broken or uprooted, "considerable structural damage"
11	103~117 (56~63)	Violent Storm	Exceptionally high waves (11.5~15m), sea completely covered with long white patches of foam	Very rarely experienced, "widespread damage"
12	118~ (64~)	Hurricane	Exceptionally high waves (14m~) with the air filled with foam and spray, sea completely white with driving spray; visibility very seriously affected	-

Wind chill index

Sensation of warm or cold is not exactly proportional to temperature. The transfer rate of heat between the skin and the ambient air depends on the wind speed. As the wind speed increases, the skin loses more heat due to the enhanced convection associated with the micro-scale turbulence. The wind chill index, equivalent to temperature in reflecting the cold sensation in calm air, is used to measure coldness in winter as it combines the effects of wind and temperature. With a wind chill temperature of -22℃, exposed skin can freeze in 15 minutes in Table 1.3.2.

Table 1.3.2. Wind chill index, an equivalent measure of sensation of coldness in units of temperature that depends on the atmospheric parameters of air temperature and wind speed. Shading indicates temperatures at which frostbite can occur. As individual responses to exposure to cold air vary, the table should be used as a rough guide. The table is reproduced from Osczevski and Bluestein (2005).

Air Temperature(℃)

Wind Speed (km/hr) \ Calm	10	5	0	-5	-10	-15	-20	-25	-30	-35	-40	-45	-50
10	9	3	-3	9	-15	-21	-27	-33	-39	-45	-51	-57	-63
15	8	2	-4	-11	-17	-23	-29	-35	-41	-48	-54	-60	-66
20	7	1	-5	-12	-18	-24	-31	-37	-43	-49	-56	-62	-68
25	7	1	-6	-12	-19	-25	-32	-38	-45	-51	-57	-64	-70
30	7	0	-7	-13	-19	-26	-33	-39	-46	-52	-59	-65	-72
35	6	0	-7	-14	-20	-27	-33	-40	-47	-53	-60	-66	-73
40	6	-1	-7	-14	-21	-27	-34	-41	-48	-54	-61	-68	-74
45	6	-1	-8	-15	-21	-28	-35	-42	-48	-55	-62	-69	-75
50	6	-1	-8	-15	-22	-29	-35	-42	-49	-56	-63	-70	-76
55	5	-2	-9	-15	-22	-29	-36	-43	-50	-57	-63	-70	-77
60	5	-2	-9	-16	-23	-30	-37	-43	-50	-57	-64	-71	-78
70	5	-2	-9	-16	-23	-30	-37	-44	-51	-59	-66	-73	-80
80	4	-3	-10	-17	-24	-31	-38	-45	-52	-60	-67	-74	-81

Heat index

Likewise, various combinations of air temperature and humidity have been converted into a heat index in units of equivalent temperature reflecting the warm sensation felt in dry air. One may get more stress from heat and feel less pleasant in summer, as humidity increases with a fixed temperature. The greater the increase in humidity, the less latent body heat is transferred to ambient air. Many national meteorological centers provide both heat index and wind chill index along with the air temperature. Heat cramps or heat exhaustion are possible for 32~46°C. Heat stroke is possible for 47~72°C. Heat stroke is very likely for 72°C and up. Exposure to intense sunlight can increase values up to 8 degrees in Table 1.3.3.

Table 1.3.3. Heat index for assessing the potential severity of heat stress. As individual responses to exposure to heat vary, the table should be used as a rough guide. The table is reproduced from http://www.princeton.edu/~oa/files/heatindx.pdf

Environmental Temperature (℃)

	12°	17°	22°	27°	32°	37°	42°	47°	52°	57°	62°
Relative Humidity	Apparent Temperature (℃)										
0%	6°	11°	15°	20°	25°	29°	33°	37°	41°	45°	49°
10%	7°	12°	17°	22°	27°	32°	37°	42°	47°	53°	58°
20%	8°	14°	19°	24°	29°	35°	41°	47°	54°	62°	72°
30%	9°	15°	20°	26°	32°	38°	46°	55°	65°	77°	90°
40%	10°	16°	21°	28°	35°	43°	52°	65°	79°	93°	
50%	11°	17°	23°	30°	38°	49°	62°	77°	92°		
60%	12°	18°	24°	32°	42°	56°	74°	91°			
70%	12°	19°	27°	35°	48°	66°	86°				
80%	13°	20°	28°	39°	55°	78°					
90%	13°	21°	30°	44°	64°						
100%	14°	22°	33°	50°							

1.4 Environmental monitoring

Air quality

One can find real-time air-quality information on the web. Exposure to pollen and fungal spores can also lead to allergic rhinitis or asthma. Some centers provide pollen counts on the web. The upper panel of Fig. 1.4.1 shows an example from the website of U.S. Environmental Protection Agency (EPA). The website provides the density of pollen with diameters less than 2.5μm, which are based on $PM_{2.5}$ measurements. $PM_{2.5}$ above 100 values is unhealthy for people who are sensitive to pollen and such allergens. Values above 200 are very unhealthy. Due to moderate wind circulation in eastern States brought by the Hurricane Ike and in the northern plains by the approaching mid-latitude systems, most of the States have moderate to high air quality.

The lower panel of Fig. 1.4.1 shows an example from the KMA website. The website provides the density of dust particles with diameters less than 10μm, which are based on PM_{10} measurements. The information is useful in monitoring the threat of the Yellow sands or Asian yellow dusts transported from the desert areas upstream of Korean Peninsula.

Many centers provide standardized indicators of air quality that measure ground level amounts of ozone (O_3), sulphur dioxide (SO_2), nitrogen dioxide (NO_2), carbon monoxide (CO), and other particulates. Chronic heart or respiratory illnesses may be observed if one is exposed to high concentrations of these pollutants or particulates. Each center uses different definitions for the air quality index, and it is necessary to check the health implications of the indices.

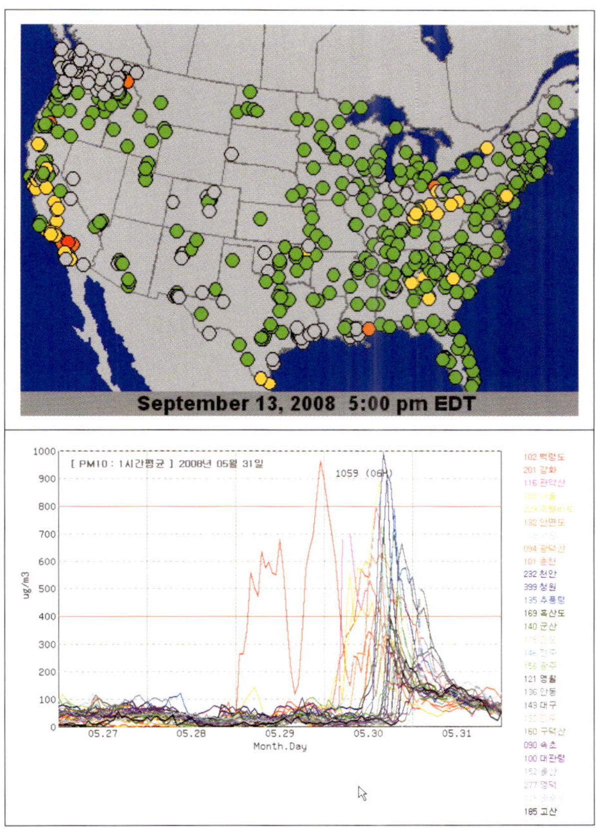

Fig. 1.4.1. (Upper) Example of air quality map for dust particle distribution in units of g/m^3 based on $PM_{2.5}$ measurement from AIRnow, U.S. Environmental Protection Agency website at 08 UTC 13 September 2008, when Hurricane Ike made landfall over Texas. Colors represent the following values: green - below 50; yellow - 50~100; orange - 100~150; red - 150~200; purple - above 200. (Lower) Another example is taken from the website of KMA in units of g/m^3 based on PM_{10} measurements. KMA issues warnings for PM_{10} value above 500 g/m^3. PM_{10} refers to particles of 10 micrometers or less in aerodynamic diameter. Both images are excerpted from http://www.aaaai.org/nab/index.cfm?p=allergenreport&stationid=134&datecount=08%2F16%2F2007 and http://web.kma.go.kr/eng/weather/asiandust/timeseries.jsp respectively.

Ultraviolet rays

Over exposure to the sun's ultraviolet (UV) rays with wavelengths ranging from 290 to 400 nanometers causes health problems including skin cancer, premature aging of the skin, cataracts and other eye damage, and immune system suppression.

Shorter wavelengths of radiation cause more damage to health. For instance, 290 nanometer radiation is three times more dangerous than 350 nanometer radiation, and five times more than 400 nanometer radiation.

The UV index is a forecast of the probable intensity of skin damaging ultraviolet radiation reaching the surface during the solar noon hour. This index takes into account the damage levels varying with the depth of stratospheric ozone layer, wavelength of UV, elevation and cloud cover. Those inputs mostly rely on satellite-measured instruments, and the quality of UV index is exposed to errors from the associated observation. More UV radiation reaches ground at higher altitudes with fewer clouds, and in more upright angles with respect to the surface. Table 1.4.1 shows the chart for checking the possibility of acquiring sunburn versus the UV index. Most national meteorological centers provide the forecast for the UV index on the web. For more detailed instruction on using the UV index, please refer the websites of NOAA and BoM:

http://www.cpc.ncep.noaa.gov/products/stratosphere/uv_index/uv_information.shtml

http://www.bom.gov.au/uv/

Table 1.4.1. UV index and protective action, as reproduced from the website of Centers for Disease Control and Prevention (http://www.cdc.gov). The UV index represents the probable intensity of skin damaging ultraviolet radiation reaching the surface during the solar noon hour. The sensitivity of the skin from the UV radiation varies from person to person. Sun protection factor (SPF) is a measure of UVB protection that ranges from 1 to 45 or above. A sun screen with an SPF of 15 filters 92% of the UVB.

Exposure Category	UV Index	Protective Actions
Minimal	0, 1, 2	Apply skin protection factor (SPF) 15 sun screen.
Low	3, 4	SPF 15 & protective clothing (hat)
Moderate	5, 6	SPF 15, protective clothing, and UV-A&B sun glasses.
High	7, 8, 9	SPF 15, protective clothing, sun glasses and make attempts to avoid the sun between 10am to 4pm.
Very High	10+	SPF 15, protective clothing, sun glasses and avoid being in the sun between 10am to 4pm.

2. Severe Weather Warnings and Forecasts

2.1 High impact weather

Weather forecasts and warnings appear first at the main page of any national meteorological center's website. Just enter the name of a city of interest at any portal site and hundreds of websites are shown to provide local forecasts and warnings. Weather forecasts usually give predictions on daily details for 7 days for the precipitation amount, maximum and minimum temperatures, cloudiness, humidity, wind speed and direction. The extended forecasts appear on a separate menu bar and give climatological conditions in probabilistic terms for the next week, coming month or season. The WWIS website of WMO links to the official weather forecasts for major cities around the world as updated daily by the national meteorological centers. They generally include daily maximum and minimum temperatures, brief textual description of the expected weather during the next few days.

In case of imminent storms, warnings are highlighted in the headline to draw the attention of the readers by using a pop-up window or a blinking banner. The pre-alerting of potential high impact weather can be made as early as 7 days ahead in probabilistic terms that reflect differences of computer model forecasts. The warnings are updated as frequently as needed to reflect the latest change in atmospheric conditions. While the weather forecast from different websites may differ with each other, the severe weather warnings are shared nationwide but under the control of national meteorological centers. The warnings available at any local site are updated as frequently as possible by the local government, avoiding any possible conflict or confusion with those from private service providers and neighboring national meteorological centers.

Heavy precipitation and strong winds are commonly observed in many parts of the world, though their physical mechanisms differ from region to region. Other severe weather phenomena such as dust storms and downslope winds are found only in specific regions. The WMO homepage provides links to the latest severe weather reports valid for the domain of each responsible national meteorological center, including thunderstorms, heavy rain or snow, and tropical cyclones. A global report comes out first on the screen after selection of the type of severe weather from the main window of the Severe Weather Information Center (SWIC) website. Clicking on the weather symbol of interest in the report will provide a more detailed weather report, as demonstrated in Fig. 2.1.1. For instance, one may access the detailed tropical cyclone warnings and track forecasts issued by the official authority with a click on the tropical cyclone symbol. The Network of European Meteorological Service (EUMETNET) website also provides a comprehensive report on the latest severe weather warnings over Europe in graphical format with color codes.

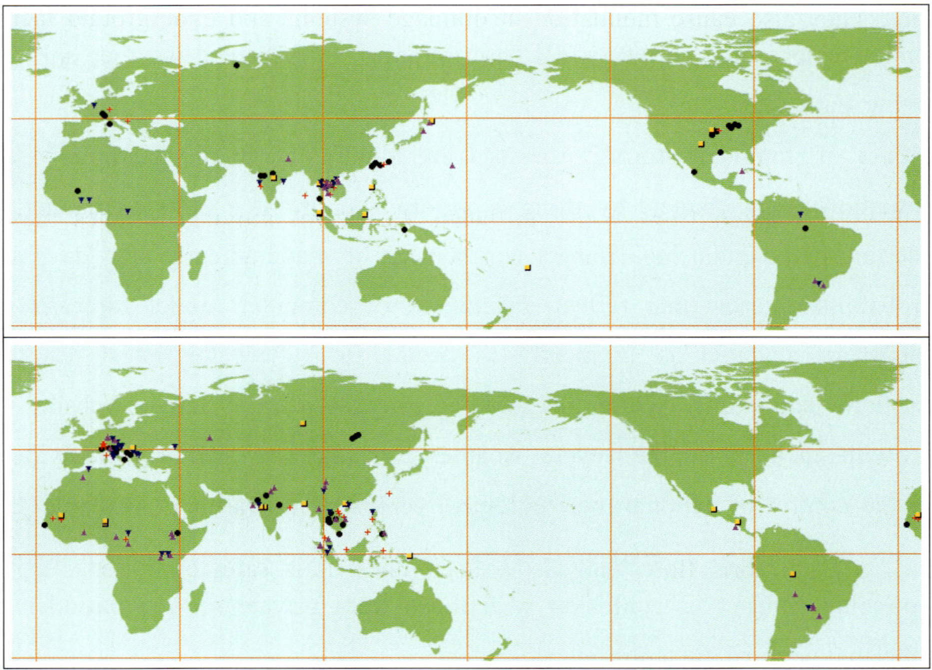

Fig. 2.1.1. (Upper) Heavy rain and (Lower) thunderstorm activity around the globe from the WMO

website at 12 UTC on 13 September 2008. Dot: reported at the latest synoptic hour; cross: 6 hours earlier; square: 12 hours earlier; inverted triangle: 18 hours earlier and triangle: 24 hours earlier. The images are available from http://severe.worldweather.wmo.int/rain/ and http://severe.worldweather.wmo.int/thunder/ respectively. Tropical cyclones and other weather information can be found at the same website.

The definition of high impact weather differs from region to region, reflecting the natural climate and socio-economic conditions. The rest of this chapter deals with the practical meaning of forecast and warning signals for the common types of severe weather phenomena such as heavy rain and snow, thunderstorm, and tropical cyclones.

2.2 Heavy rain and snow

Heavy rains lead to flooding in low-lying areas, and landslides on mountainous regions. They also cause inundation in drainage systems and underground facilities in urban areas, as floating debris blocks roads and bridges. In general, hourly rainfall of more than 40 mm threatens lives of city dwellers, and campers in the valleys. The damage from rainfall primarily depends on its intensity, as illustrated in Table 2.2.1 taken from a city in Japan. Breaks of moving vehicles may not work properly when rainfall rates exceed 30 mm per hour. It would be very dangerous to drive when rainfall rates exceed 50 mm per hour. Disastrous situations may arise when rainfall rates exceed 80 mm per hour (http://www.jma.go.jp/jma/kishou/know/yougo_hp/amehyo.html). If your car stalls due to flooding, abandon it immediately and climb on to higher ground. Stay away from downed power lines and electric wires. In addition, if you are in areas susceptible to landslides and debris flow, consider evacuation if it is safe to do so.

Table 2.2.1. Impact of concentrated heavy rain, as illustrated for the city of Osaka Japan (http://www.city.suita.osaka.jp/kakuka/bousai/PDF_English/p14-15.pdf).

Rainfall rate (mm/hr)	Impact
5~10	Puddles are formed immediately, and the clear sound of raindrops is heard.
10~20	A sheet of rain blankets the ground. Hard to hear speaking voice due to the sound of raindrops.
20~30	A downpour. Potential overflow of drains and landslides may occur.
30 ~	Heavy rain as if a bucket of water was turned over. An evacuation order may be issued at a danger area.

In winter, heavy snowfall brings havoc to local communities through the deformation of structures and facilities, and the failure of electrical grids and transportation, resulting in a shortage of goods and other products required for daily living. Few individuals have developed scales that link heavy snowfall to its social impact. A scale called the Northeast Snowfall Impact Scale (NESIS) in Table 2.2.2 reflects the snowfall impact mostly on Northeastern United States, which may be applicable to other countries. Other winter storm scales can be found in Cerruti and Decker (2011). Map symbols for the precipitation intensity are presented in the Table 2.2.3.

Table 2.2.2. Impact of heavy snowfall, as illustrated for the northeastern United States. The Northeast Snowfall Impact Scale or NESIS scale considers the amount of snow, the size of the area covered and the population residing in that area to estimate the human and economic impact of the storm. The table is reproduced from Kocin and Uccellini (2004).

NESIS Index	Impact
Notable (1~2.499)	Large areas of 10 cm (4 inch) accumulations and small areas of 25 cm (10 inch) snowfall.
Significant (2.5~3.99)	Significant areas of greater than 25 cm (10 inch) snows while some include small areas of 50 cm (20 inch) snowfalls.

NESIS Index	Impact
Major (4~5.99)	Large areas of 25 cm (10 inch) snows (generally between 80 and 240 × $10^3 km^2$, roughly 1~3 times the size of the state of New York) with significant areas of 50 cm (20 inch) accumulations.
Crippling (6~9.99)	Huge areas of 25 cm (10 inch) snowfalls, and each case is marked by large areas of 50 cm (20 inch) and greater snowfall accumulations. Order may be issued at a danger area.
Extreme (10.0~)	25 cm (10 inch) accumulations exceed 320 × $10^3 km^2$ and affect more than 60 million people

Table 2.2.3. Precipitation intensity for rain, drizzle, and snow for use in surface weather charts reproduced from the Appendix II-4 in the manual on the global data-processing system of WMO (http://www.wmo.int/pages/prog/www/WMOCodes/VolumeI1.html#VolumeI1)

	Slight	Moderate	Heavy
Intermittent	●	● ●	● ● ●
Continuous	●●	● ●●	● ●● ●

2.3 Lightning, hail, and strong wind

Individual thunderstorms are very localized areas of severe weather phenomena with diameters of less than 30 km that can be found in both the tropics and extratropics. Some 1,800 thunderstorms are occurring at any moment around earth. They are accompanied by lightning, hail storms, tornados, and flash floods. They bring powerful destruction to communities in their path.

Thunderstorms develop where air is vertically unstable, like the boiling water in the pot. The conditions can be met when warm and humid air comes to the lower atmosphere or when dry and cold air flows to the upper atmosphere. Typically the wedge of warm air near the cold fronts provides favorable conditions for the initiation of storms. The vertical instability depends on the profile of temperature and humidity. Meteorologists design various indices that reflect the major characteristic of the profile by assuming a simple parcel displacement or bulk type of cloud model. Some of the indices are introduced in Table 2.3.1. The vertical shear contributes to the organization of convective cells, even though it is not directly responsible for the vertical instability. The horizontal shear could turn into a tornado force wind if lifted vertically during the right conditions.

Table 2.3.1. Vertical instability index that measures the possibility of thunderstorms and potential severity. Many national meteorological centers provide the forecasts of various instability indices on the web. The threshold values in the table represent typical thunderstorms that occur across the continental United States. Different climate zones may require the adjustment of the threshold values accordingly. The table is reproduced from http://www.crh.noaa.gov/lsx/science/indices.php.

Severity	CAPE (J/Kg)	Lifted index (℃)	K index	Total Totals
Marginal	0 to 999	0 to -2	20 to 25	44 to 45
Moderate	1000 to 2500	-3 to -5	26 to 30	46 to 51
Strong	2500 to 4000	-6 to -9	31 to 35	52 to 55
Extreme	4000 and above	-9 and below	36 and above	56 and above

Thunderstorms also can occur when the ground is heated by sunlight to produce buoyancy of the air parcel near the ground. It is called an 'air mass thunderstorm' and is typical of the middle of summer or along long coastal section. The development of this type of thunderstorm is generally intermittent and sporadic, and often accompanied with showers. Map symbols for the showers are presented in the Table 2.3.2. A squall is a general name for a strong, gusty wind with rain or snow that is accompanied by a thunderstorm. The lazy S is also shown for freezing rain or freezing drizzle.

Table 2.3.2. Map symbols for squalls and showers, which are reproduced from the Appendix II-4 in the manual on the global data-processing system of WMO (http://www.wmo.int/pages/prog/www/WMOCodes/VolumeI1.html#VolumeI1).

Symbol	Description	Symbol	Description
▽	Squall		
● ▽	Rain shower	● ▽ (heavy)	Heavy rain shower
✱ ▽	Snow shower	✱ ▽ (heavy)	Heavy snow shower
(●∿	Freezing rain	(●∿●)	Heavy freezing rain

Lightning

Lightning heats the air in the path of the stroke to about 25,000℃, a value that is equivalent to keeping a 100-watt light bulb lit for more than 3 months. For every 600,000 lightning strokes, one injury occurs. It would be safe to stay indoors if you cannot count to 25 after seeing lightning and before hearing thunder. When you are outdoors during a thunderstorm, try to get into a building or car. If no structure is available, go out to find an open space and squat low on the ground as quickly as possible. Avoid natural lightning rods and tall structures. Lightning also is a major cause of wildfires.

Hail

It can be dangerous when thunderstorms produce hail that is 20 mm or larger. One estimate says that a 20 mm hailstone can fall at 65km/hr (18m/s). Figure 2.3.1 shows the impact of large hail on car damage. When hail size increases over 3 cm, the panels begin to severely dent. When the hail size exceeds 7 cm, 40% of damaged cars have their windows broken.

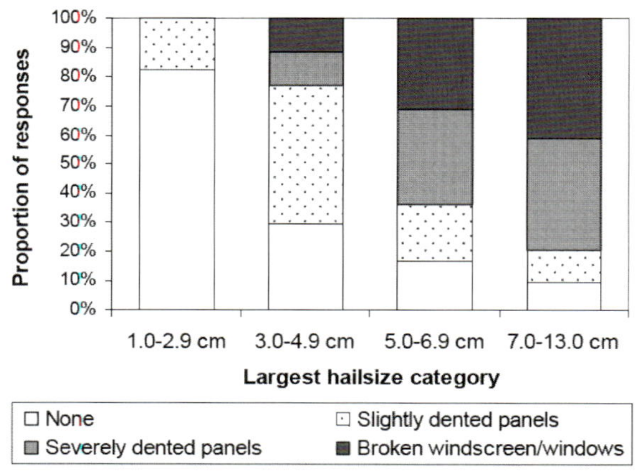

Fig. 2.3.1. Large hail size and car damage from Leigh (2007).

Strong wind

The straight-line wind or Derecho also frequently occurs during a severe thunderstorm. Tornado force winds embedded in a rotating thunderstorm ravage the lives and properties along their path and are often accompanied by lightning and hail. A thousand tornados occur every year in United States. A distant second is Canada, with around 100 tornadoes per year. Other locations that experience frequent tornado occurrences include Australia, Bangladesh, Japan, Korea, New Zealand, northern Europe, and western Asia. Most of them last a few minutes and move only a few miles. If a tornado is encountered while in an automobile, drive perpendicular to the movement of the tornado. In the outdoors, find a cave and something to hang on to, or lie flat in a ditch and cover your head. When indoors, go to the basement or the lowest floor and cover yourself with a blanket. A damage scale for the tornado force winds is presented in Table 2.3.3 as an extension of Beaufort scale. The scale can be also applied to the strong winds produced by mature extratropical and tropical cyclones. The warnings related to the tropical cyclones are discussed separately in the following section.

Table 2.3.3. The TORRO tornado intensity scale (or T-Scale) measures tornado intensity between T0 and T10. This table is reproduced from what was developed by Terence Meaden of the Tornado and Storm Research Organization (TORRO), a meteorological organization in the United Kingdom (http://www.torro.org.uk/TORRO/severeweather/tscale.php).

T-Scale	Fujita Scale	Beaufort Scale	Wind Speeds Km/hr (mph)	Tornado Description	Damage
FC			None	Funnel Cloud	No damage except to tops of extremely tall towers, balloons, and aircraft. Generally no damage on land, except to the tops of very large trees.
T0	0.1	7.70	63 ~ 87 (39 ~ 54)	Light	Loose, light weight object spiraled in air from the ground; twigs snapped and visible path through crops.
T1	0.6	11.53	88 ~ 116 (55 ~ 72)	Mild	Things like light chairs and potted plants and other slightly heavier objects become airborne. Dislodging of tiles on roofs becomes more evident. Minor damage to bushes and trees; wooden fences can be flattened.
T2	1.1	15.40	117 ~ 148 (73 ~ 92)	Moderate	Heavy mobile homes displaced, light caravans blown over, garden sheds destroyed, garage roofs torn away, much damage to tiled roofs and chimney stacks. General damage to trees, some big branches twisted or snapped off, small trees uprooted.
T3	1.6	19.20	150 ~ 183 (93 ~ 114)	Strong	Mobile homes heavily damage or possibly over turned. Light caravans may be destroyed; Garages and weaker outbuildings are destroyed as well; House roof timbers are thoroughly exposed; Bigger trees uprooted or destroyed
T4	2.2	23.10	185 ~ 219 (115 ~ 136)	Severe	Cars lifted from ground; Mobile homes become airborne and are destroyed; roofs lifted from some houses; roof timbers of stronger brick and stone houses are completely exposed; many trees uprooted or snapped in half.

T-Scale	Fujita Scale	Beaufort Scale	Wind Speeds Km/hr (mph)	Tornado Description	Damage
T5	2.7	26.90	220 ~ 257 (137 ~ 160)	Intense	Heavy cars lifted into the air; more serious house damage than that of a T4, but walls usually remain standing; some of the oldest and weakest building may collapse completely.
T6	3.2	30.80	259 ~ 299 (161 ~ 186)	Moderately Devastating	Well-built houses lose roofs completely and possibly one or two walls; many slightly weaker built building collapse completely.
T7	3.7		301 ~ 341 (187 ~ 212)	Strongly Devastating	Wooden framed houses completely destroyed; some walls built of stone or brick beaten down or collapsed; steel framed warehouse-type constructions may buckle slightly; cars thrown through the air; noticeable debarking of trees caused by flying debris.
T8	4.2		343 ~ 386 (213 ~ 240)	Severely Devastating	Cars hurled great distances; Wood framed houses and object inside are spread out across long distances; irreplaceable damage to stone or brick houses; steel framed building buckled.
T9	4.8		388 ~ 433 (241 ~ 269)	Intensely Devastating	Numerous steel framed buildings are damaged moderately; cars and trains hurled long distances; complete and thorough de-barking of any standing tree trunks.
T10	5.3		434 ~ 481 (270 ~ 299)	Super	Entire framed houses and other similar building lifted bodily from the foundation and hurled great distances; steel-reinforced buildings and such are severely damaged.

2.4 Tropical cyclones

The WMO defines a tropical cyclone by the maximum wind speed exceeding 63 km/hr (34 knot or 17 m/s) that is equivalent to Beaufort scale 8. It is noted, however, that meteorological centers use different averaging schemes to measure wind speed. WMO suggests conversion factors between 10 minute averaging and 1 minute averaging: 0.93 at sea, 0.90 at off sea, 0.87 at off land, and 0.84 inland (Harper et al., 2008).

The tropical cyclones form over deep tropical ocean bodies where the sea surface temperature is above 27℃. The eight regions prone to tropical cyclones are shown in Fig. 2.4.1. More than 70% of the tropical cyclones occur in the northern hemisphere. Specifically the northwest Pacific and Bay of Bengal are vulnerable areas with frequent passages of tropical cyclones. Each region uses different names for this dangerous weather phenomenon. They are called typhoons in the northwest of the Pacific, cyclones in South Asia and Oceania, and hurricanes in the Atlantic.

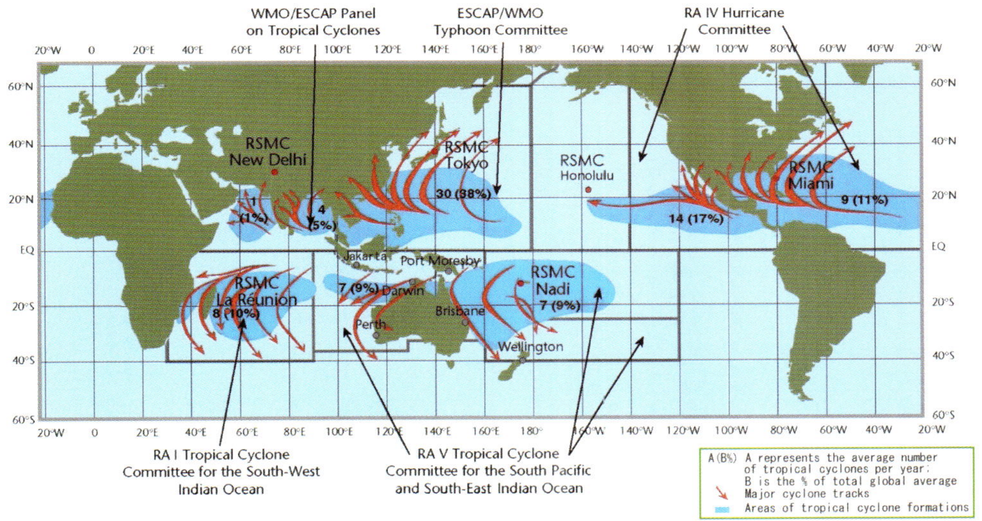

Fig. 2.4.1. Five regional organizations and frequency of passage of tropical cyclones. The regional specialized meteorological centers (RSMC) are indicated in the map. The figure is kindly provided from the Tropical Cyclone Programme (TCP) of WMO. Similar image is available from http://www.wmo.int/pages/prog/www/tcp/Advisories-RSMCs.html.

Tropical cyclones are notorious for their massive power of destruction caused by the strong winds, heavy rains, and storm surges. The wind speed at the ground, as a combined effect of the straight-line movement of the storm and the circular motion embedded in it, is greater at the right-hand side of the moving storm. The radius of 63 km/hr winds extend several hundred kilometers away from the center of circulation; this average speed is equivalent to the strength required for a pedestrian to feel difficulty in walking. Often heavy rainfall occurs in the spiral bands of thunderstorms that extend outward from the eye. Storm surges of 6 meters high water wall on top of 3 meter waves are not unusual. The combined surge causes deaths and severe damage to coastal communities lying along the path of the storm. The ocean water pulled into the storm's low pressure center is transformed into taller water pillars as it approaches the coast with shallow bathymetry. High waves induced by the gale force winds simply add to the height of the water walls. The Saffir-Simpson hurricane damage scale in Table 2.4.1 is widely used in the United States; a similar version has been developed for the Pacific region and is named the Saffir-Simpson Tropical Cyclone Scale (STiCkS). This latter scale was independently devised by Guard and Lander (1999).

Table 2.4.1. Saffir-Simpson hurricane category scale. This table is reproduced from http://www.srh.noaa.gov/srh/jetstream/tropics/tc_classification.htm.

Category	Sustained wind speed Stormsurge Central pressure	Potential
1	119~153 km/hr (74~95 mph) 1.2~1.5 m 980 hPa	No real damage to building structures. Damage primarily to unanchored mobile homes, shrubbery, and trees. Also, some coastal flooding and minor pier damage.
2	154~177 km/hr (96~110 mph) 1.8~2.4 m 965~979 hPa	Some roofing material, door, and window damage. Considerable damage to vegetation, mobile homes, etc. Flooding damages piers and small craft in unprotected anchorages may break their moorings.
3	178~209 km/hr (110~130 mph) 2.7~3.7 m 945~964 hPa	Some structural damage to small residences and utility buildings, with a minor amount of curtain wall failures. Mobile homes are destroyed. Flooding near the coast destroys smaller structures with larger structures damaged by floating debris. Terrain may be flooded well inland.

Category	Sustained wind speed Stormsurge Central pressure	Potential
4	210~249 km/hr (131~155 mph) 4.0~5.5 m 920~944 hPa	More extensive curtain wall failures with some complete roof structure failure on small residences. Major erosion of beach areas. Terrain may be flooded well inland.
5	≥250 km/hr (≥156 mph) ≥5.5 m <920 hPa	Complete roof failure on many residences and industrial buildings. Some complete building failures with small utility buildings blown over or away. Flooding causes major damage to lower floors of all structures near the shoreline. Massive evacuation of residential areas may be required

The latest forecast and warnings on tropical cyclones can be found on the websites of the specialized regional meteorological centers. Tropical storm symbols are presented in Table 2.4.2. The WMO website links to the national meteorological centers that are responsible for the issuance of tropical cyclone warnings at local level.

Table 2.4.2. Tropical storm symbols which are reproduced from the Appendix II-4 in the manual on the global data-processing system of WMO.

	Northern hemisphere	Southern hemisphere	Criteria
Tropical depression	○	○	Below 33 kt
Tropical storm			33~64 kt (63~119 km/hr)
Tropical cyclone			Above 64 kt

The impact of a tropical cyclone is very sensitive on the location and timing of its landfall. Small departures between the forecasted track and the observed track could cause major changes in disaster prevention measures for communities under the greatest risk. Thus, the major meteorological centers introduced probabilistic concepts for the issuance of forecast and warnings, examples of which are presented in Figs. 2.4.2 and 2.4.3. More details on the probabilistic forecasts will be discussed in Chapter 6.

The forecasted tracks of tropical cyclones are expressed in terms of probability circles exceeding a certain threshold value, or the chance that the forecasted center of tropical cyclone will be located inside of the circle. Many centers use 60% or 70% as the threshold value. The intensity of the storm is inversely proportional to the central pressure. The outer boundary of gale force wind is defined in terms of the radius of wind speed of 63 km/hr (17 m/s or 34 knot). In many cases, the storms have asymmetric wind structures; so does the radius of influence.

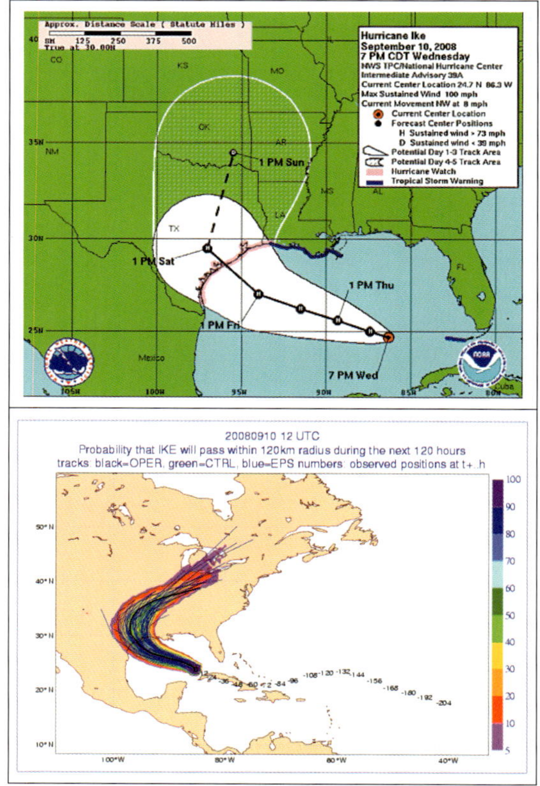

Fig. 2.4.2. Examples of track forecasts from the U.S. National Hurricane Center, NHC (upper) for D+3

forecast starting from at 00 UTC on 11 September 2008 with range of potential threat, and the European Center for Medium-range Weather Forecasts, ECMWF (lower) for D+5 forecast starting from 12UTC on 10 September 2008 with probability of striking at any locality. The images are available from http://www.nhc.noaa.gov and http://www.ecmwf.int/products/forecasts/d/tccurrent respectively.

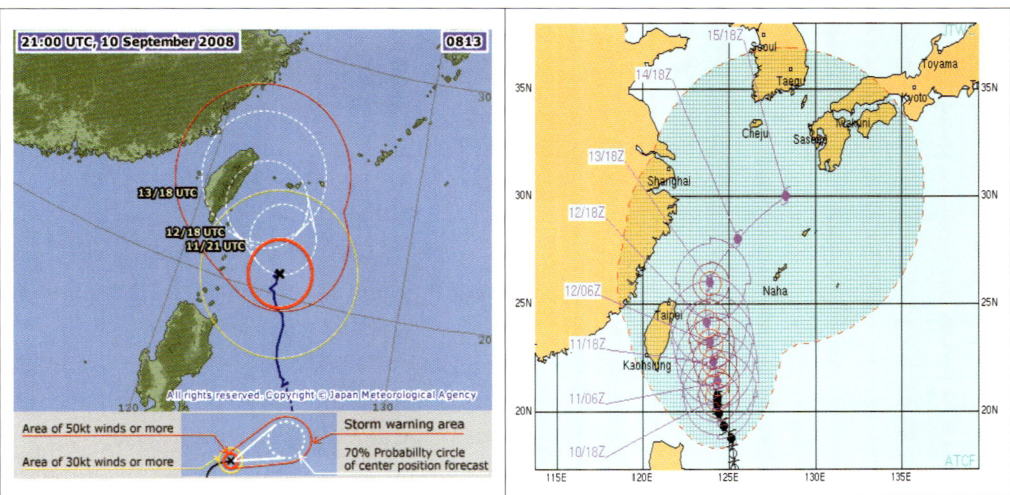

Fig. 2.4.3. Examples for the track and/ or intensity forecasts of the typhoon Sinlaku from JMA (left) and Joint Typhoon Warning Center, JTWC (right), issued at 12 UTC on 10 September 2008. The uncertainty of track forecasts is represented by the probability circle (JMA) or ship avoidance areas (elongated lines in JMA and JTWC). The radius in JTWC forecast represents 63 km/hr or 34 kt winds (outermost), 93 km/hr or 50 kt winds (middle), and 119 km/hr or 64 kt winds (innermost). The radii in JMA forecast represent 50 kt winds with 30 kt winds only at the starting point. The wind speeds for the ship avoidance area are 93 km/hr (50 kt) for JMA, 63 km/hr (34 kt) for JTWC respectively. The images are available from http://www.jma.go.jp/en/typh/ and http://www.usno.navy.mil/JTWC respectively. Similar information can be found in many websites of national meteorological centers.

Table 2.4.3. Major centers that issue tropical cyclone warnings. More details are found at the tropical cyclone programme, WMO. (http://www.wmo.int/pages/prog/www/tcp/Advisories-RSMCs.html)

Area covered	URL (http://)
Caribbean Sea, Gulf of Mexico, North Atlantic and eastern North Pacific Oceans	RSMC Miami-Hurricane Center/NOAA/NWS National Hurricane Center, USA www.nhc.noaa.gov/index.shtml

Area covered	URL (http://)
Western North Pacific Ocean and South China Sea	RSMC Tokyo-Typhoon Center/Japan Meteorological Agency www.jma.go.jp/en/typh/
Bay of Bengal and the Arabian Sea	RSMC-tropical cyclones New Delhi/India Meteorological Department www.imd.gov.in
South-West Indian Ocean	RSMC La Réunion-Tropical Cyclone Centre/Météo-France www.meteo.fr/temps/domtom/La_Reunion/
South-West Pacific Ocean	RSMC Nadi-Tropical Cyclone Centre/Fiji Meteorological Service www.met.gov.fj/advisories.html
Central North Pacific Ocean	RSMC Honolulu-Hurricane Center/NOAA/NWS, USA www.prh.noaa.gov/hnl/cphc/
South-East Indian Ocean	TCWC-Perth/Bureau of Meteorology (Western Australia region), Australia www.bom.gov.au/weather/wa/cyclone/
Arafura Sea and the Gulf of Carpenteria	TCWC-Darwin/Bureau of Meteorology, Australia www.bom.gov.au/weather/cyclone/
Coral Sea	TCWC-Brisbane/Bureau of Meteorology, Australia www.bom.gov.au/weather/cyclone/
Solomon Sea and Gulf of Papua	TCWC-Port Moresby/National Weather Service, Papua New Guinea
Tasman Sea	TCWC-Wellington/Meteorological Service of New Zealand, Ltd. www.metservice.co.nz/forecasts/severe_weather.asp

Part Ⅱ. Forecast, Verification and Probability

3. Forecast Process

4. Understanding Weather Charts

5. Quality of Weather Information

6. Probability and Application

3. Forecast Process

3.1 Standard forecasting procedure

The weather forecast is the combined product of human judgment with the computer-based forecast chart. The standard forecasting procedure for the practical weather forecasting is well summarized by Coiffier (2004): (1) understanding the present meteorological situation, (2) examination of the pertinence of the analysis, (3) identification of the key elements of the meteorological situation, according to conceptual models accepted by the forecast community, (4) examination of the various computer forecasts or ensemble products and the user's choice of the most likely scenario, (5) description of the evolution of the atmosphere corresponding to the chosen scenario, (6) deducing the consequences for smaller scales and specific areas, (7) description of the expected weather in terms of weather elements, (8) decision about the opportunity to issue special warning, and (9) transmitting the various products to specific users.

Forecasters use a diversity of approaches for the stages (1) to (3); as a result, the initial procedures are difficult to standardize. The initial decisions at this stage are subjective. Additional discussion on these stages is detailed in section 5.4. Users may enjoy the application of forecast products without knowing the thought processes that occur in stages (1) to (3). Various computer model forecasts and analyses are used heavily in the stages (2) and (5) among others. Section 3.2 discusses the computer forecast and the concepts used by numerical models on the cyberspace. High-resolution model guidance and other statistical guidance aid in the process in stages (6) and (7). Chapters 9 to 13 deal with the interpretation of high-resolution model products and downscaling procedures. The warnings that might be warranted at stage (8) are extensively introduced in Chapter 2, where users learn to convert the message into action words in taking precautions against any imminent weather hazards.

In general, users are content with the final forecast decisions and end-state products issued at stages (8) and (9). However, to appreciate the limitation and uncertainty of forecast information, one should pay attention on to the background information forecasters produce at intermediate stages. The background information also helps strengthen the confidence about the forecast. Part of background information for the areal forecast is discussed throughout the rest of this section, while section 3.3 focuses on the point forecasts produced at stages from (7) to (9).

The U.S. National Weather Service (NWS) make readily available on their website the prognostic reasoning behind the official forecast and severe weather warnings. An example is presented in Fig. 3.1.1. It contains the diagnosis of atmospheric conditions such as vertical instability and shear, pre-existing boundaries, and other local features that are conducive to storm development. A discussion follows on satellite images, radar signatures, and other mesoscale observations, in conjunction with the interpretation of the latest mesoscale model runs and their limitations.

MESOSCALE DISCUSSION 2263
NWS STORM PREDICTION CENTER NORMAN OK 0251 PM CDT SAT SEP 13 2008
AREAS AFFECTED…CENTRAL/ERN OH…SWRN PA…WV PANHANDLE CONCERNING…SEVERE POTENTIAL…WATCH POSSIBLE VALID 31951z - 132145z.

CONVECTION WAS BEGINNING TO DEVELOP ALONG WARM FRONT FROM CENTRALINTO SERN OH. AIRMASS WAS MARGINALLY UNSTABLE WITH MUCAPES FROM 1000-2000 J/KG. MODERATELY LOW LEVEL SHEAR MAY SUPPORT THE POTENTIAL FOR ISOLATED TORNADOES WITH ANY DISCRETE CELLS. AREA IS BEING MONITORED FOR POSSIBLE WW.

SFC ANALYSIS SHOWS A SLOW MOVING WARM FRONT EXTENDING FROM NEAR MANSFIELD OHIO SEWD TO NEAR ZANESVILLE THEN EWD ACROSS THE WV PANHANDLE AND FAR SWRN PA. A FEW CONVECTIVE CELLS HAVE DEVELOPED ALONG AND JUST NORTH OF THIS BOUNDARY IN THE WAA REGIME. PRESENCE OF MODERATE LOW-MID LEVEL VERTICAL SHEAR OWING TO S-SELY SFC WINDS/ COMBINED WITH MODERATE INSTABILITY MAY FAVOR DISCRETE TSTMS CAPABLE OF PRODUCING ISOLATED TORNADOES /ESPECIALLY WITH ANY CELL THAT CAN SUSTAIN A DEVIANT STORM MOTION INVOF FRONTAL BOUNDARY/.

..CROSBIE.. 09/13/2008

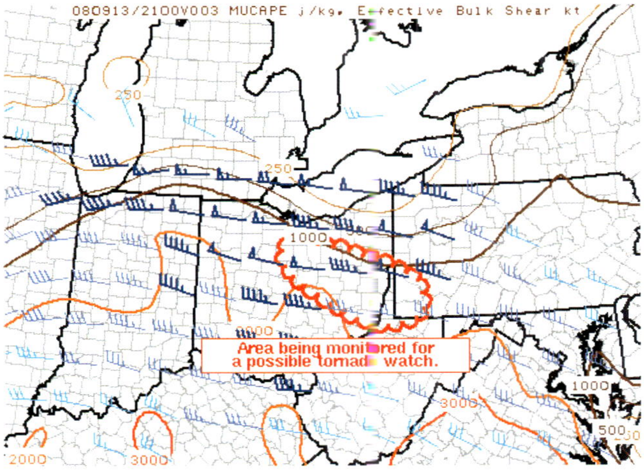

Fig. 3.1.1. Example for reasoning in support of the mesoscale forecast issued at 05 UTC on 13 September 2008 (upper), and possible tornado watch (lower) at south of the stationary front near Great Lakes. The images are available from http://www.spc.noaa.gov/products/md/. Source: U.S. NWS Storm Prediction Center, SPC.

Similar background information can be found for tropical cyclone warnings issued at JTWC, NOAA, and JMA. It generally starts with the analysis of position fixing and the structure of the tropical cyclone using satellite images and other auxiliary observations. Then follows a discussion on prognostic reasoning behind forecasts of tracks and intensities, based on model predictions and statistical guidance. An example from Joint Typhoon Warning Center, JTWC is presented in Fig. 3.1.2.

WDPN31 PGTW 101500
MSGID/GENADMIN/NAVMARFCSTCEN PEARL HARBOR HI/JTWC//
SUBJ/PROGNOSTIC REASONING FOR TYPHOON 15W WARNING NR 09//
RMKS/
1. FOR METEOROLOGISTS.
2. 12 HOUR SUMMARY AND ANALYSIS
A. TYPHOON 15W HAS INTENSIFIED 20 KNOTS OVER THE PAST 12 HOURS WITHIN A VERY FAVORABLE ENVIRONMENT. …… DYNAMIC MODEL AIDS ARE IN MUCH BETTER AGREEMENT. HOWEVER THERE REMAINS A FAIR AMOUNT OF SPREAD IN TRACK DEPICTION AFTER TAU 24, AND SEVERAL OUTLIERS EXIST WITH EGRR TO THE EXTREME WEST AND TCLAPS AND GFDN TO THE EAST OF THE CONSENSUS.
B. TY 15W IS TRACKING GENERALLY NORTHWARD UNDER THE STEERING INFLUENCE OF THE MID-LEVEL SUBTROPICAL RIDGE (STR) POSITIONED NORTHEAST OF THE SYSTEM, BUT IS ALSO BEING INFLUENCED SLIGHTLY BY A FINGER OF THE STR POSITIONED NORTHWEST OF THE SYSTEM, WHICH IS SERVING TO LIMIT FORWARD TRACK SPEED. …… THERE IS EXCELLENT CONFIDENCE IN THE CURRENT POSITION AND STORM MOTION WHICH ARE BASED ON SATELLITE FIXES, SUPPORTED BY INFRARED IMAGERY AND A 100956Z SSMI IMAGE THAT DEPICT A WELL DEFINED EYE FEATURE. ……

3. FORECAST REASONING
A. SINCE THE PREVIOUS ISSUANCE OF THE PROGNOSTIC REASONING, THE FORECAST TRACK HAS BEEN SHIFTED WESTWARD DUE TO ENHANCED INFLUENCE OF THE STR POSITIONED NORTHWEST OF THE SYSTEM. ……
B. …… THE SYSTEM IS FORECAST TO ACCELERATE AHEAD OF AN APPROACHING MIDLATITUDE SHORTWAVE TROUGH WHICH IS TRACKING EASTWARD ACROSS CENTRAL CHINA. THE 101200Z 500 MB ANALYSIS INDICATES THAT THIS APPROACH IS A RELATIVELY DEEP, DYNAMIC TROUGH WITH SIGNIFICANT HEIGHT FALLS OVER EASTERN CHINA. THE MODELS ARE IN GOOD AGREEMENT ON TRACKING THE SHORTWAVE INTO THE EAST CHINA SEA BY TAU 24, WHICH SHOULD WEAKEN THE STR NORTHWEST OF THE SYSTEM AND ALLOW FOR AN INCREASE IN FORWARD TRACK SPEED AS THE STORM BEGINS A SLOW NORTHEASTWARD TURN. TY 15W IS FORECAST TO CONTINUE INTENSIFYING THROUGH TAU 48 DUE TO EXCELLENT RADIAL FLOW, WARM SST AND HIGH OCEAN HEAT CONTENT. ……
C. IN THE EXTENDED PERIOD, THE SYSTEM IS FORECAST TO ACCELERATE AND TRACK MORE NORTHEASTWARD AND IS EXPECTED TO GRADUALLY WEAKEN AS IT INTERACTS WITH THE BAROCLINIC ZONE BEGINNING NEAR TAU 72, THE EXTENDED MODEL GUIDANCE IS IN POOR AGREEMENT.

Fig. 3.1.2. Meteorological reasoning behind the typhoon forecast issued at 12 UTC on 10 September 2008 from Joint Typhoon Warning Center, JTWC in conjunction with the track forecast in Fig. 2.4.3. This type of information is available from http://www.usno.navy.mil/JTWC.

As Internet-based broadcasts become popular, some websites provide video casting to explain the reasoning behind the weather forecast and to help users understand the scientific background. This approach is similar to what one can see when a weather caster explains the weather forecast for today and tomorrow on the Weather Channel. As conditions warrant, Internet based broadcasting and its contents can be updated on a real-time basis to reflect its substantial advantage during rapid changes in the atmospheric state. For instance, the KMA website (http://www.weather.kr/weatherinfo/explanation.jsp) provides the expert's comment on current weather and forecasts along with the interpretation of various images of satellite and radar, and available computer forecast charts.

This behind-the-scenes information is written in plain terminology so users can understand the uncertainty and limitations associated with weather forecasting, without contending with the jargon of meteorology. Understanding how weather forecasts are produced gives the users the idea that various forecast scenarios are equally possible and that some of them might go wrong for different reasons. It also helps these individuals appreciate the probabilistic character of forecasts, and the expected risk of forecast failure in each scenario.

3.2 Computer model forecast

Computer forecasts, as discussed in the previous section, are produced through a sequence of scientific processes, i.e., observation, communication and collection of data, quality control, numerical analysis and prediction, post-processing for graphics, verification and applications. The standardization of the data-processing with computer facilities is a well-established practice under WMO, as clearly illustrated in the manual on the global data-processing and forecasting system (GDPFS).

The GDPFS deals with the hierarchy of information infrastructure among national meteorological services. Broadly speaking, this infrastructure has three levels: global center, regional center and national center. Each national meteorological service can be classified into one of these centers in terms of the capacity and supporting infrastructure for weather and climate forecast service. The global centers lead the cutting-edge science and training, and support the main hub for the global dissemination of their forecasts. The regional centers play a similar role as global center but on the regional scale. The national center is responsible for service across the national territory with the support of global or regional centers. The global centers listed in Table 3.2.1 provide model predictions on the global scale for distribution to partners in GDPFS. The Observing System

Research and Predictability Experiment (THORPEX) program and associated demonstration projects, designed to enhance the early warning capabilities of national meteorological services, are based on this concept to facilitate the cascading flow of information among the three level partners.

Table 3.2.1. Basic characteristic of operational global numerical weather prediction systems as of 2010, mostly based on the Working Group on Numerical Experimentation (WGNE), jointly established by the World Climate Research Program (WCRP) Joint Scientific Committee and the WMO Commission for Atmospheric Sciences (CAS). L: number of vertical levels; M: number of ensemble members; T: Triangular truncation; TL: Triangular with Linear reduced Gaussian grid; N: wave number. In practical sense, T, TL, and N indicate number of waves a global model can resolve. WGNE has the responsibility to foster the development of atmospheric circulation models for use in weather prediction and climate studies on all time scales and for diagnosing and resolving shortcomings. The table is available from http://wcrp.wmo.int/documents/wgne25rpt.pdf.

Forecast Centre (Country)	Medium-range prediction Forecast-range	Ensemble prediction Forecast-range
CAWCR, BoM (Australia)	80km L50 10 days	TL119 (108km) L19; M33 10 days
CMA (China)	TL639 (20km) L60 10 days	T213 (60km) L31; M15 10 days
CMC (Canada)	0.45°x0.3° (16km) L58 10 days	0.9° (32km) L28; / M20 16 days
CPTEC/INPE (Brazil)	T299 (43km) L64 7 days	T126 (102km) L28; M15 15 days
DWD (Germany)	30km L60 7 days	
ECMWF (Europe)	TL1279 (10km) L91 10 days	TL639 (20km) L62 10 days
HMC (Russia)	T85 (150km) L31 10 days; 0.72°x0.9° (26km) L28 10 days	
JMA (Japan)	TL959 (13km) L60 9 days	TL319 (40km) L60; M51 9 days
KMA (Korea)	N512 (25km) L70 10 days	N320 (40km) L70; M24 10 days
UKMet (U.K.)	N512 (25km) L70 6 days	N216 (60km) L38; M24 3 days

Forecast Centre (Country)	Medium-range prediction Forecast-range	Ensemble prediction Forecast-range
Météo France (France)	TL538 (24km) L60 4 days	TL358 (36km) L55; M35 4 days
NCEP (U.S.A.)	TL511 (25km) L80 7.5 days; TL254 (50km) L80 7.5~16 days	T126 (102km) L28; M45 16 days
NCMRWF (India)	T254 (50km) L64 7 days	T80 (160km) L18; M8 7 days
NRL/ Navy (U.S.A.)	T239 (54km) L30 7.5 days	T119 (108km) L30; M16 10.5 days

The atmospheric state can be observed from various platforms including satellites, aircrafts, hot air balloons and land and marine-based stations. The atmosphere constantly changes along three-dimensional spaces. As such, the atmosphere requires a huge amount of data to describe itself. The smoke from cigarettes in a small room, a miniature of the atmospheric motion field, can not be captured even with a modern digital camera. The volume of atmosphere on earth is 10^{16} times greater than that of a typical smoking room. One can imagine how much pixel data is required to describe motion in the global atmosphere.

The atmosphere is reconstructed in cyberspace where its motion is simulated with the support of computers. The computerized model of the atmosphere in the cyberspace consists of many small boxes, each of which represents a unit volume of atmospheric state to be specified by temperature, wind, pressure, and humidity. The state of each box is affected by the state of the adjacent box through algebraic operations, such as addition and multiplication, that must follow the scientific laws. By chain reaction, the box on one side of the globe eventually has some connection to the remote boxes located on the opposite side. The number of data boxes used to describe the atmospheric state at an advanced meteorological center is on the order of 10^{6}.

Raw observations from various platforms differ in terms of variables, area coverage, frequency, and time. For a given time, these observations are merged, synchronized, converted to common variables, and finally transferred onto a regularly-spaced array of grid points. For instance, the radiance from satellite or reflectivity from radar has to be converted to humidity and temperature profiles, or vice versa. The sophisticated data assimilation schemes incorporate the variational technique known as 4-dimension variation data assimilation (4dVar). This technique is used in the complex analysis that is essential to numerical weather prediction.

Once various types of observation are fed into the numerical model, the model projects into the future and produces forecast charts for 7 days or more. The model is a system of software that draws on governing scientific equations of the atmosphere. It computes the standard output variables such as mean sea level pressure, relative humidity, dewpoint temperature, and geopotential height. It also derives special field variables, including vertical instability indices, tropopause-level data, flight-level fields, freezing-level information and boundary-layer fields. Table 3.2.2 illustrates various outputs used by forecasters. The procedure for the numerical weather prediction is well documented in WMO (1993).

Table 3.2.2. List of typical model output variables as presented in U.S. National Center for Atmospheric Research, NCAR community model and ECMWF model. The table is reconstructed from the information available from Persson (2001) and https://www.eol.ucar.edu/projects/gcip/dm/documents/map_99_lsa_nw/appendix_b.html.

	Variables
Surface	Mean sea level pressure Wind vector at 10 m Maximum wind gust at 10m 2-meter specific humidity (or dew point) 2-meter temperature 2-meter maximum and minimum temperature Skin temperature Soil temperature (all soil layers) Soil moisture (all soil layers) Latent heat flux (surface evaporation)

	Variables
Surface	Sensible heat flux Ground heat flux Surface momentum flux Snow depth (water equivalent) Snow melt Surface runoff Sub-surface runoff Surface downward short-wave radiation flux Surface upward short-wave radiation flux (gives albedo) Surface downward longwave radiation flux Surface upward longwave radiation flux
Single level	Tropopause height (or pressure)
Vertical level	Geopotential height Temperature Specific humidity Wind vector Moisture flux vector Omega (vertical motion, - Dp/Dt) Absolute or potential vorticity Divergence Convective precipitation latent heating rate Stable precipitation latent heating rate Shortwave radiation latent heating rate Longwave radiation latent heating rate Cloud ice/ water content Cloud fraction
Vertical sum or average	Total precipitation Large-scale precipitation Convective precipitation Snowfall Total cloud cover Low, medium, high and convective cloud cover Total column water vapor Convective Available Potential Energy (CAPE) Top-of-atmosphere net longwave radiative flux Top-of-atmosphere net shortwave radiative flux

	Variables
Metadata Fixed Fields	Terrain height Roughness length Max soil moisture capacity Soil type Vegetation type

The prediction of a future state of the global atmosphere using a model requires high speed computing resources. Some centers produce global forecasts in real time basis; they are presented in Table 3.2.1, and some examples are presented in Chapter 4.

Most of the websites only permit access to the final outcome of the numerical model. However, some centers allow running a model in an interactive manner. For instance, U.S. NOAA Air Research Laboratory (ARL) enables the user to interactively run the trajectory model, the so-called the HYbrid Single-Particle Lagrangian Integrated Trajectory model or HYSPLIT, and receive the forecast of a trajectory or dispersion of the target population given that the user specifies options such as the wind input, source location, projection hours, and output intervals. An example is given in Fig. 3.2.1. The same model can be used to obtain the trajectory of dust storms, ashes from wild fires or volcanic eruption, chemical spills, and nuclear pollutants.

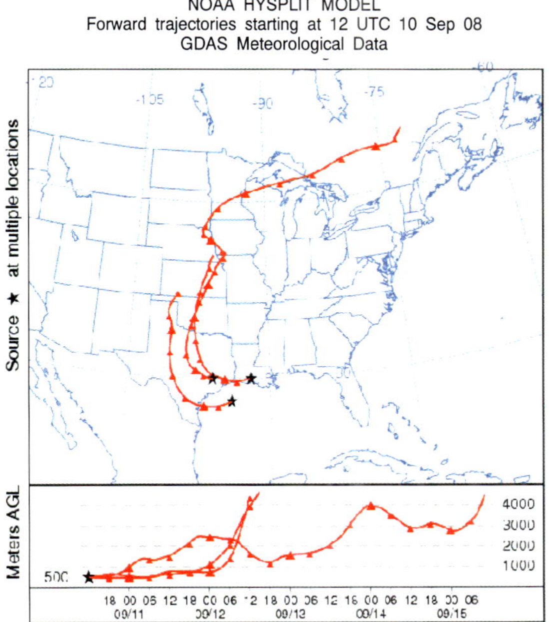

Fig. 3.2.1. Trajectories of source material on the map (upper), and vertical cross-section along the parcel trajectory (lower) from interactive run of the HYSPLIT model, U.S. NOAA Air Research Laboratory with initial condition at 12 UTC on 10 September 2008. Three-point sources near Texas spread out to the northeast, while drifting upward during a 5-day forecast period, as Hurricane Ike approached Texas before recurving to the northeastern seaboard. The image is available from http://ready.arl.noaa.gov/HYSPLIT_traj.php.

3.3 Point forecast and digital database

Typical weather forecasts describe the basic weather elements at 6~12 hour time intervals and at widely-spaced districts covering major cities and its suburbs. An example is the following: "calm and cloudy this evening, and occasional showers tomorrow in Seoul and Koungki Province, Korea". The format is suitable for mass media such as newspaper, radio and television dealing with the general public as a user group. But this text format cannot meet the users' specific needs at any particular location and special time of the day.

On the other hand, numerical forecasts are produced at narrowly-spaced grid points that are less than a few tens of kilometers apart, and for very short time intervals such as minutes. The basic weather elements are forecasted several days ahead in an advanced computer center. These numerical weather prediction forecasts, in their base format, are simply a database containing digital numbers associated with regularly spaced grid points; it is at these grid point locations that the arithmetic operation of a forecast model occurs. The numerical database includes the forecast of basic variables like pressure, temperature, wind and humidity, and specific variables listed in Table 3.2.2. Some numerical models evaluate prognostic variables on a staggered array of grid points. Spatial interpolation on a common grid is conducted before the forecast data are visualized on charts. In principle, one can forecast the weather with the spatial and temporal resolution allowed by the computer model.

An example of time series from a computer model is shown in Fig. 3.3.1 for cloud coverage, 6-hr accumulated precipitation, wind speed, and temperature from the ECMWF model. The precipitation, wind speed and cloudiness at Tokyo increased on the 19 September 2008 with the arrival of the typhoon Sinlaku, which already mentioned in Fig. 2.4.2. Many meteorological centers provide the time-series forecast for weather elements at a specific location that are directly driven from the model output. Users should note that the meteogram for any point is based on the interpolation from the surrounding model grids, typically several tens of kilometers apart. The errors associated with this interpolation should be considered when one desires to know forecast details by reading the meteograms. A coastal city maybe surrounded by different types of model grids: grid points located over land versus those located over ocean. Because the climate over land is quite different from the climate over a large body of water, one should check beforehand the exact grid point to which the meteogram refers.

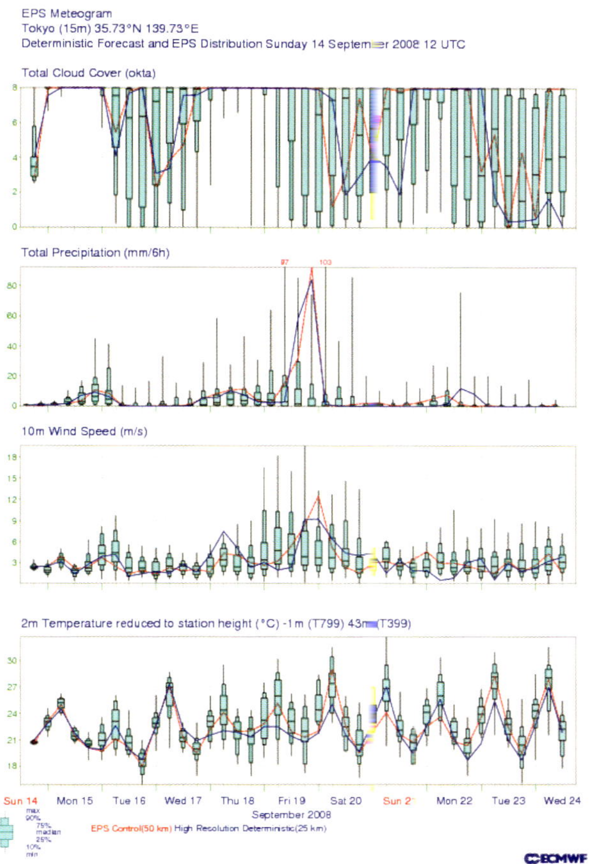

Fig. 3.3.1. Meteograms for Tokyo, Japan from an ensemble prediction system at ECMWF that produced forecasts for cloud cover, precipitation, wind speed at 10m above ground level, and temperature at 2m; the initial conditions were from 12 UTC on 14 September 2008. The uncertainty of forecasts for the various weather elements are expressed by either the rectangular box around each forecast and by the 'Whiskers' with the 25th and 75th percentiles for the wide boxes and the 10th to 90th percentiles for the narrower boxes. The ensemble model predicted the rainfall on 19 September when the weakened typhoon Sinlaku was expected to pass around Tokyo as illustrated in Fig. 2.4.2. The image is available from http://www.ecmwf.int/products/forecasts/d/charts.

Traditionally, the three-dimensional grid point values in the model output have been visualized on two-dimensional charts, and are disseminated to the media by graphical form of transmission. As two-way communication technology advances, users could directly access digital values produced by the computer for each grid point in the forecast array.

Some national meteorological centers including KMA and U.S. NWS now provide local forecasts for specific points through the digital forecast database, as shown in Fig. 3.3.2. A digital forecast is a four-dimensional array that consists of numerical values of weather elements at every grid point, each representing geographical areas as small as 5km×5km for a given time interval. Sensible weather elements are included such as extreme temperature, wind gustiness, precipitation amount, snowfall amount, relative humidity, cloud cover, and wave height. The database is updated every 1~3 hours. The graphic weather forecast is provided in terms of contours and color fills on the geographic information system (GIS) background map. One can access the detailed local forecast by dragging the cursor on the GIS map for the specific location. Click on any point of the GIS map from the main menu to receive a time series forecast for various weather elements, such as the temperature, probability of precipitation, precipitation amount, sky conditions, wind, and humidity. A multimedia data-handling professional can download the grid point values directly from the database for further application in a standard data format such as the gridded binary code, GRIB2. Those formats are designed to convey grid data in two or three dimensions. Some users may receive grid data on a regular, pre-arranged schedule by contract.

The difference of digital forecasts from the conventional ones is best understood by using an analogy of comparing digital photographs with film photographs. The digital camera is different from the traditional one in that it allows one to save the image in a digital form, and to retrieve it as many times as you want. The photo taken by the digital camera is nothing but a set of binary numbers each representing color pixels on a two-dimensional grid. It can be read by the computer. One could easily edit the digital photograph and combine it with music or media files to create multimedia art.

By the same token, the digital forecast can be read and edited through the platform of a personal computer or mobile phone, while the traditional forecasts would not. The emergence of digital forecasts opens the new application field employing multimedia technology. For example, assume we produce an animation of rainfall distribution during the next 48 hours. Many years ago, one had to acquire weather charts in 6 hourly intervals, then write specific symbols on the chart, and fill colors for the area of interest. To produce animation, one should draw similar charts repeatedly by extrapolating them forward in time. Last, one should scan the sequence of charts to convert them in digital form. Much labor is required to produce animated weather charts. Using the digital forecast database, one can easily apply the graphic tools, for instance Adobe Photoshop, to make copies, to generate 2-dimensional contours, to fill colors, and to edit the digital image at will in the personal computer environment.

The model is a simplified version of the real atmosphere. Observed conditions in the model crudely represent the true atmospheric state. The model atmosphere deviates more from nature as the forecast lead-time increases. Frequent forecast updates become necessary to correct the forecast back to an approximate of the real atmosphere. This process is analogous to the restoration of an old painting through the digital image processing technique. The first step is to reconstruct the image of the painting onto a set of digital values on fine mesh grids. The second step is to correct the digital values contaminated by the noise, then convert them back into the original image of the painting. The only difference is that the correction of the digital forecast transpires in four-dimensional space including time, not in two dimensions.

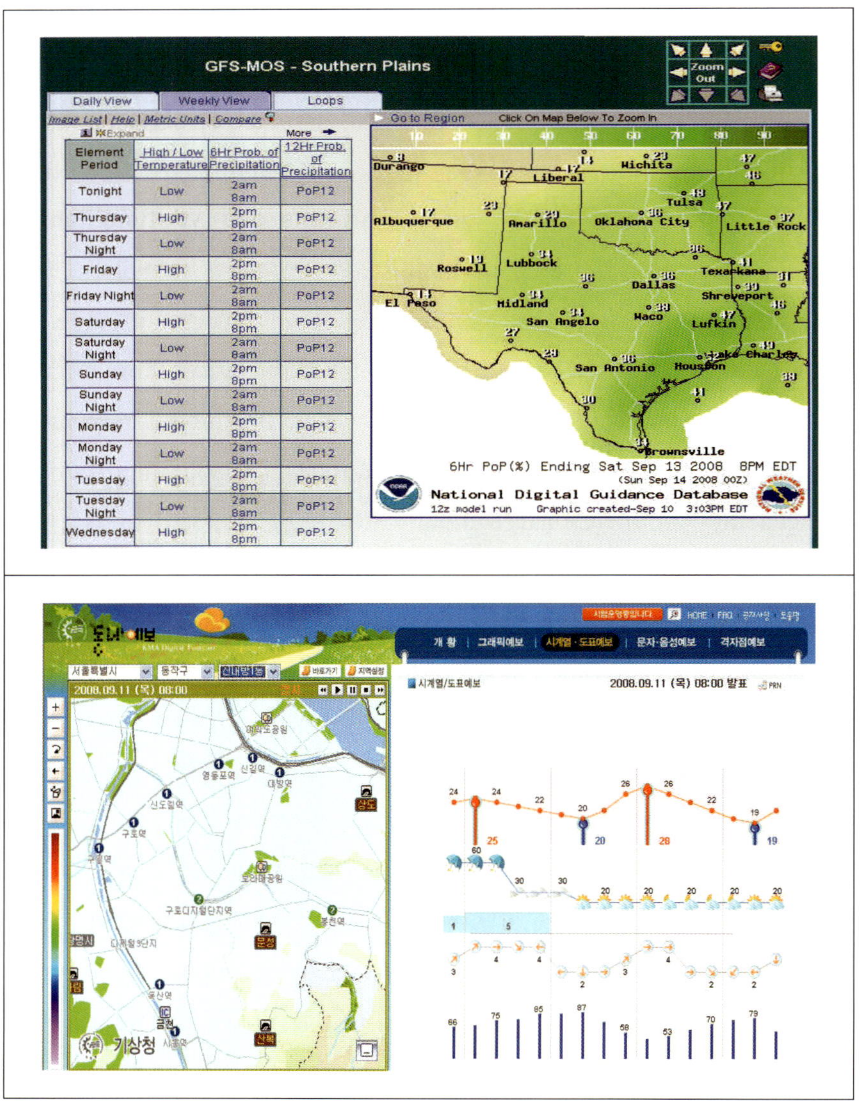

Element Period	High / Low Temperature	6Hr Prob. of Precipitation	12Hr Prob. of Precipitation
Tonight	Low	2am 8am	PoP12
Thursday	High	2pm 8pm	PoP12
Thursday Night	Low	2am 8am	PoP12
Friday	High	2pm 8pm	PoP12
Friday Night	Low	2am 8am	PoP12
Saturday	High	2pm 8pm	PoP12
Saturday Night	Low	2am 8am	PoP12
Sunday	High	2pm 8pm	PoP12
Sunday Night	Low	2am 8am	PoP12
Monday	High	2pm 8pm	PoP12
Monday Night	Low	2am 8am	PoP12
Tuesday	High	2pm 8pm	PoP12
Tuesday Night	Low	2am 8am	PoP12
Wednesday	High	2pm 8pm	PoP12

Fig. 3.3.2. Examples of digital forecasts as displayed in websites of national meteorological centers. U.S. NWS (upper) for an 84-hour projection from the initial conditions at 00 UTC on 10 September 2008, and KMA (lower) for 24-hour projection from initial conditions of 00 UTC on 11 September 2008. The images are available from http://www.weather.gov/forecasts/graphical/sectors/maryland.php#tabs and http://web.kma.go.kr/eng/weather/forecast/timeseries.jsp respectively.

Reviewing the interactive process to update the digital forecast database helps understand how to use them. At first, temperature, wind, humidity and precipitation amounts are predicted using the computer model at each 5km x 5km grid mesh, based on physical principles and a simple interpolation technique. The model forecast contains systematic bias due to the imperfection of the model physics and initial conditions. One could assess the bias and correct them by the statistical analysis of departure between the model forecasts and verifying analysis over many simulation runs. Various weather elements are corrected in this way, such as the sky condition, probability of precipitation, and maximum and minimum temperature.

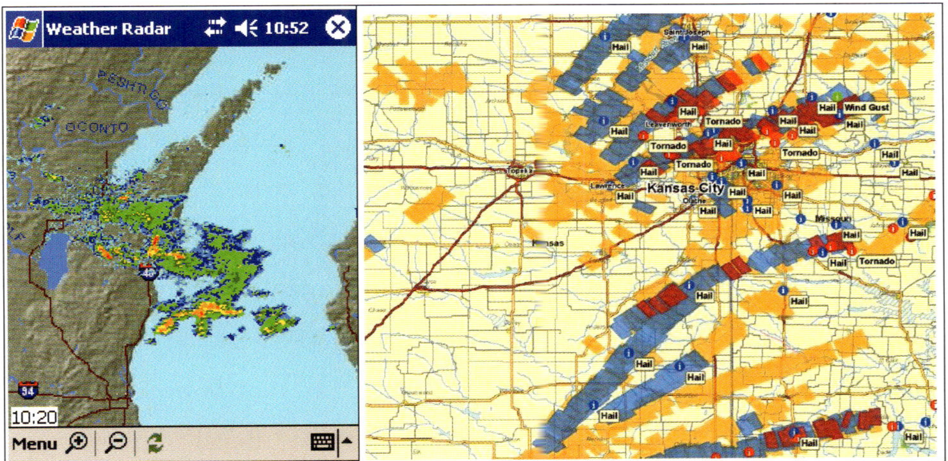

Fig. 3.3.3. Examples for the application of geographic information system (GIS) on weather forecast (Sznaider, 2005). Radar image is viewed on mobile phone (left). Time-series roll-ups of past weather activity can show the geographical footprint of areas recently affected by specific weather phenomena (right). The images are available from http://www.memory-map.com/weather_radar.htm and http://www.meteorlogix.com/pdf/MxInsight/METEORLOGIX_MXI_WHTPAPER.pdf respectively.

Finally, the forecaster fills the gap between the model and nature by correcting the digital forecast database with reference to local knowledge and experience, as the numerical models have certain limitation in representing local details characterized by mountains and valleys, land-sea contrast, and source regions nearby for moisture or pollutants. The forecasters play a vital role in extreme weather conditions when computer forecasts and statistical guidance are not capable of the task at hand. Many centers independently developed interactive software to modify the digital forecast database and to visualize the contents with multimedia tools. Some examples for digital forecasts are shown in Fig. 3.3.2.

The reliability of digital forecasts depends on the performance of the numerical models, effectiveness of statistical interpretation tools, and the expertise of forecasters. Digital forecasts have higher spatial and temporal resolution than do conventional forecasts. As information content increases in the digital forecast database, it is more likely that the accuracy of the forecast would tend to decline. Despite the high reliability of conventional forecasts compared to digital forecasts, the latter provide more fine-scale information compared to the former. For the time being, it is safe to use both the conventional forecast and digital forecast, until the digital forecasts have proven themselves to be as accurate as conventional forecasts. In addition, various application tools need to be further developed to increase the user-friendliness of digital forecast databases.

The digital forecast database is not fully utilized until it is explicitly used as input into subsequent applied models in various areas including air quality and dust monitoring, tracing chemical spills, and predicting various health indices. For example, consider a case for the application of the digital forecasts on the transport problem. When nuclear accidents or chemical spills occur, one would want to forecast the path and density of pollutants at the downstream region. To do so, one may access the digital wind forecasts from the Internet, feed that information into an application module, and run the trajectory module with one's personal computer. Thousands of potential application areas lie in waiting for use in various disciplines including agriculture, construction, leisure and travel, fisheries, etc. Examples for the application of digital forecasts are shown in Fig. 3.3.3.

4. Understanding Weather Charts

4.1 Why read weather charts

Driving to a resort for the first time, John asked a bystander how to get there. "Find exit 21, then turn left. After finding the sign for Resort D, go straight ahead for about 30 minutes." John missed the exit thinking about the business contract he'd received at the office. He had to park the car by the roadside, and looked at the map to see where he was. Soon he found the turnaround to get back to the highway. He managed to keep on the right track, as long as he was looking at the map.

Interpreting weather charts is similar to the map reading in the above example. Let's consider the following forecast for a certain county: "rain showers will begin this afternoon, and will stop tomorrow." What can one do if the rain actually comes in the evening instead of the afternoon? Should one dismiss the forecast? The same steps taken by the driver who adjusted his route using a map would apply to local weather forecasts. The weather chart would take the place of the road map in this case. If one can read a weather chart, one could figure out the gross features of the atmospheric conditions and changes therein. Even when the point forecast does not perfectly match the current weather situation, one can revise the forecast to accommodate local changes in the weather. This example demonstrates the usefulness of the weather chart in judging the soundness of the forecast, and in building confidence for applications.

The weather chart describes the state of atmosphere in detail as it constantly evolves through time. Forecasts and warnings are based on the weather chart, a working sheet used by forecasters, as discussed in section 3.1. The homepage of any major meteorological center features many weather charts with varying heights, variables, and valid times. Weather charts fall into two categories, one for the analysis of present or past conditions, and the other for the forecast chart for the appropriate forecast time range. Each part can be further divided into originals produced by the computer and versions edited by a weather specialist with fronts and other weather symbols. Various weather charts are available free of charge from the Internet. For instance, ECMWF, KMA, JMA, and NCEP provide online weather charts for various atmospheric variables at 2.5km, 5km, 7km, and 8.5km height surfaces above ground. More detailed discussion will be given later on how to interpret weather charts.

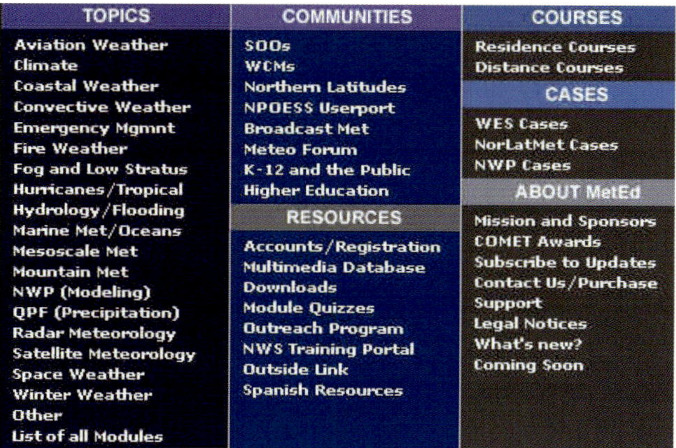

Fig. 4.1.1. A list of subjects appeared on the meteorology education and training website of University Cooperation for Atmospheric Research, UCAR (www.meted.ucar.edu) where many self-learning modules created by the Cooperative Program for Operational Meteorology, Education and Training, COMET, can be found.

There are thousands of textbooks that explain how to read weather charts. A variety of computer-aided learning modules have also been developed to assist users in interpreting forecast charts. The website of Cooperative Program for Operational Meteorology, Education and Training (COMET) in Fig. 4.1.1 is a good example, providing overviews of subjects including nowcasting, mesoscale analysis, aviation weather forecasting, numerical weather prediction, and satellite image interpretation. Other organizations or training schools for practical weather forecasting include the U.K. Met-Office College and Ecole Nationale de la Météorologie. This chapter only introduces the very basics for using weather charts; serious readers are encouraged to visit the websites listed in Table 4.1.1 for further study.

Table 4.1.1. Some websites with training materials for weather forecasting, developed with the computer-aided learning (CAL) technique (Coiffier, 2004).

Name	Subject	Address (http://)
ANASYG-PRESYG	Computer aided learning focused on synoptic meteorology from the point of view of the forecaster	www.meteorologie.eu.org/anasyg
ASMET	Computer aided learning focused on forecaster's work in Africa, developed by EAMAC (Niger) and Eumetsat	training.eumetsat.int/course/category.php?id=35
COMET	Very comprehensive set of modules covering meteorological phenomena and weather prediction techniques	www.comet.ucar.edu/
EUROMET	Course on NWP and satellite meteorology, developed by a consortium of meteorological services and universities	www.eoportal.org/directory/info_EuroMETEuropeanMeteorologicalEducationandTraining.html
SATREP	Methodology devoted to the interpretation of meteorological information by the forecaster, developed by ZAMG, FMI, KNMI and EUMETSAT	www.knmi.nl/satrep/

The state of the atmosphere can be described using the distribution of surface pressure, wind, temperature and humidity. Air motion in the atmosphere is bounded to roughly 10km above the surface. As the pressure gradually decreases with altitude, pressure is often used as the vertical coordinate. The pressure at a given level is the weight of the atmosphere. It is measured in units of hectopascals (hPa or 10^3 Nm^{-2}). 1000 hPa is equivalent to the weight of a 10m water column.

The vertical resolution of a numerical model, as presented by the number of levels to the model top, determines the degree of vertical complexity in the model atmosphere. Many models have grid intervals varying in height with dense representation near the surface. Models with more levels have a better representation of the vertical structure of the atmosphere including the boundary layer. A model with limited vertical resolution has difficulty representing explicitly shallow layers of stratocumulus and fine features of the boundary layer. Figure 4.1.2 illustrates an example from the U.K. Met Office Unified Model (UM) with 38 levels. The model grossly captures the observed profile, but misses details in the vertical wind shear and lapse rate. The wind departures from observation is evident near surface and the layer between 300~400 hPa. Particularly, the surface southerlies and adiabatic mixed layer between 925~1000 hPa are not present in the model sounding. The model generally depicts the saturation of moisture near the surface and 400 hPa, but poorly represents the dry zone in the layer between 700~500 hPa and the saturation in the layer between 200~400 hPa.

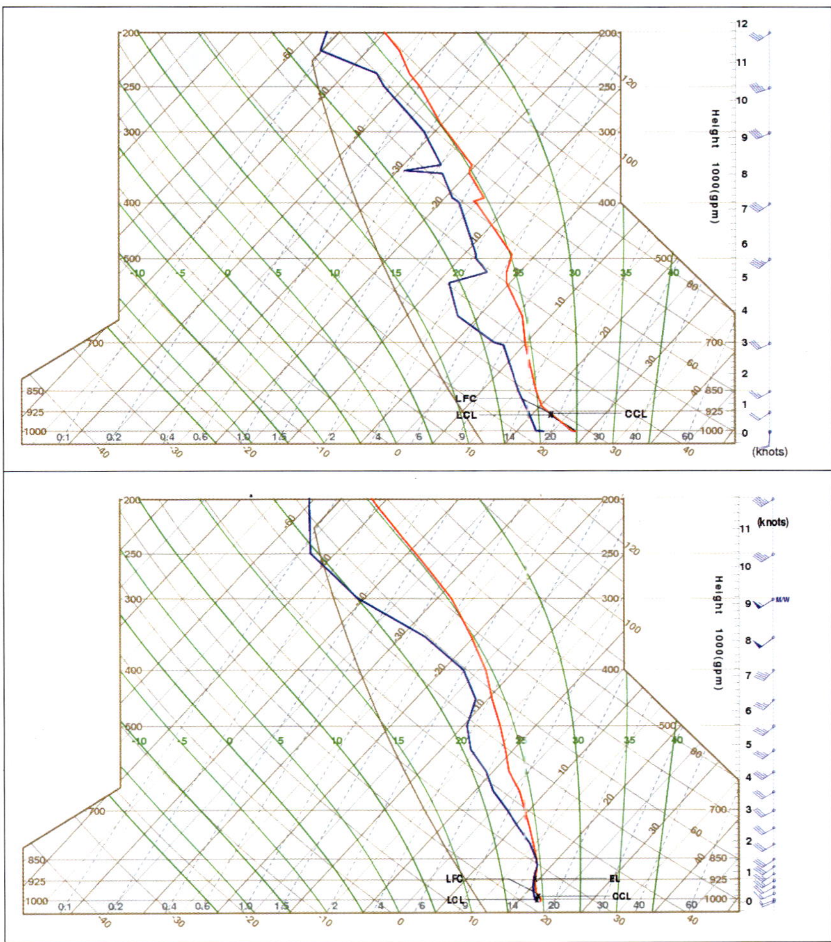

Fig. 4.1.2. Thermodynamic diagram near Seoul at 15 LST on 10 September 2010, before the passage of a cold front. The observation (upper) is compared with the model simulation (lower). The UM model with a grid spacing of 12km and 38 levels was used for the simulation. Source: KMA.

4.2 Mid-level charts

The three-dimensional state of the atmosphere can be viewed from a number of two-dimensional surfaces that cut through the atmosphere in parallel with the surface. The mean character of atmospheric motion can be represented on weather charts in the middle of the troposphere, which corresponds to 500 hPa in pressure coordinates, or roughly 5km above the mean sea level. The 500 hPa chart is useful in assessing the horizontal distribution of heat and momentum, and in finding out where warm or cold air masses are located and in which direction they are moving along with westerlies. The information provided by 500 hPa charts has further implications on the evolution of weather systems near the surface.

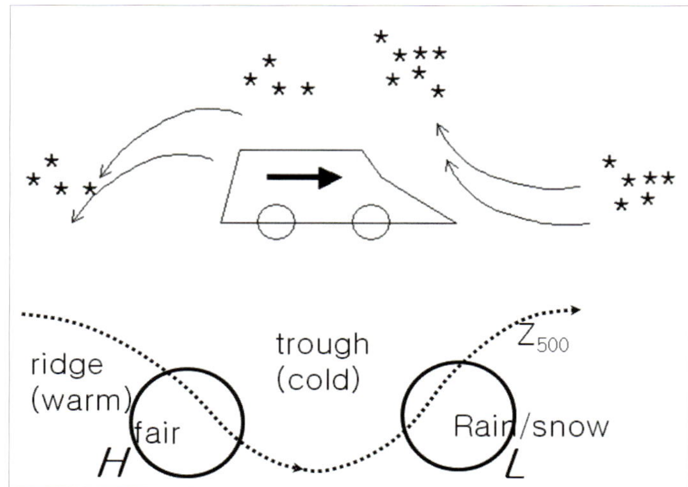

Fig. 4.2.1. Analogy between weather systems and snowplows. A cold air mass sits under the upper level trough; and a warm air mass is located beneath the upper-level ridge (lower). A surface low develops downstream of the upper-level trough, and a surface high builds up behind the same trough in a mature weather system. Lifting of warmer air in front of the upper-level trough induces clouds and precipitation, while the sinking of colder and drier air suppresses them on the rear. This state of affairs is similar to the effect of a snowplow. As the snowplow moves forward, fresh snow comes in front and old snow is dumped out behind. Likewise fresh air comes aloft to form clouds and precipitation in front of the upper trough; and old air escapes out the rear of the same trough (upper). Unsettled weather persists during the passage of a surface low, and retreats with the approach of a surface high.

The height of the 500 hPa surface is proportional to the mean temperature of the lower atmosphere. The cold dome in the lower troposphere corresponds to the trough at 500 hPa. Likewise the warm core in the lower troposphere is found beneath the ridge at 500 hPa. The deeper the trough is at 500 hPa, the colder is the region beneath. Thus a warm or cold spell near the ground depends on the phase and intensity of the ridge or the trough at 500 hPa.

In the midlatitudes, the weather system is often tall enough that the advancing trough at 500 hPa is closely tied to the surface low downstream or to the surface high upstream. A weather system functions very much like a snowplow in a snow field, as illustrated in Fig. 4.2.1. The boundary-layer air in front of the storm is fed into the surface low, forced upward to grow into deep clouds, and spreads out along the upper current as mobile clouds. The upper air in the rear of the storm falls down to the surface, whose forward branch spreads out beneath the warm air to support the upward motion in front, fueling the storm. It is noted that the air kept in the system is constantly changing, i.e., new air comes in front of the system and replaces the old air, is transformed into different phases of hydrometeors, and is finally displaced at the rear of the system. The ridge and trough at 500 hPa normally move eastward in midlatitudes. At the same time, they meander north and south in a wave pattern. Their propagation speeds depend on their curvatures. A closed low moves more slowly than an open trough.

Let's take the example of the ECMWF analysis chart in Fig. 4.2.2, and then move on to the NCEP and ECMWF forecast charts in Figs. 4.2.3 and 4.2.4. Both figures were valid at 12 UTC on 13 September 2008. The analysis of geopotential at 500 hPa in Fig. 4.2.2 shows well-developed troughs at 30°W, 55°W, 115°W, 165°W, 140°E, 100°E and series of cut-off lows over Eastern Europe extending to the Middle East in the northern hemisphere. Severe weather in recent hours over Eastern Europe and northeastern states of America in Fig. 2.1.1 is located downstream of midlatitude troughs. In the southern hemisphere, there are six troughs; the one at 50°W corresponds to the severe weather that occurs near the southern part of Brazil. The thermal trough at 850 hPa, shown in color, was generally in phase with the geopotential trough at 500 hPa. The cold air mass from the polar region extends southward where the geopotential trough developed in mid-troposphere. It is noted that the vast area of convective weather in Central Africa, South Asia and Central America in Fig. 2.1.1 can hardly be captured in the upper-level chart except for Hurricane Ike near Texas. Additional analysis on vertical instability and other local information is required to diagnose and forecast tropical weather.

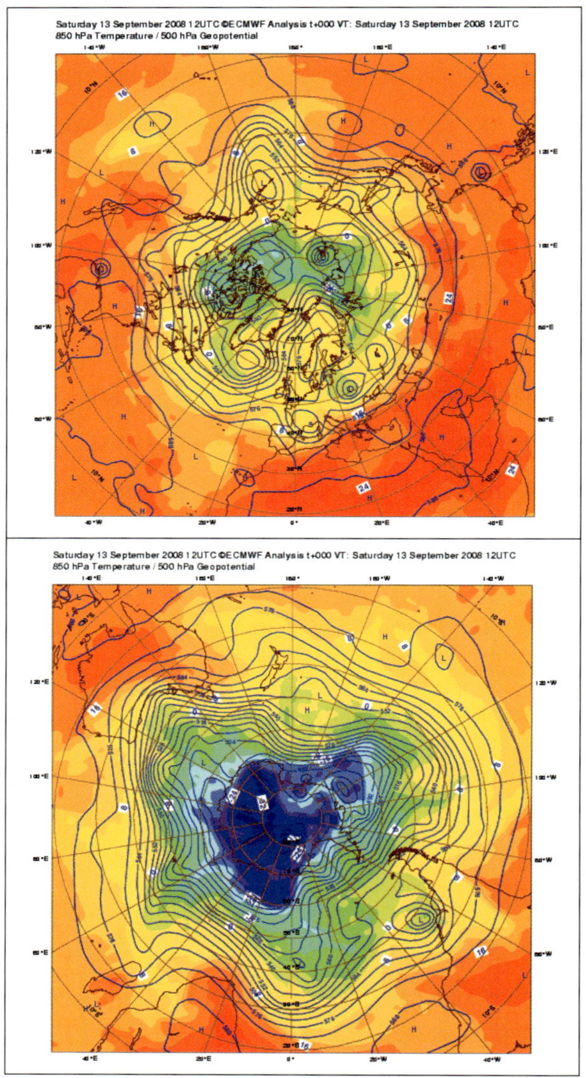

Fig. 4.2.2. Geopotential (solid line) at 500 hPa and temperature (color fill) at 850 hPa from ECMWF analysis at 12 UTC on 13 September 2008. The upper and lower panels represent the northern and southern hemisphere respectively. The images are available from http://www.ecmwf.int/products/forecasts/d/charts/medium/deterministic.

The analysis chart for 500 hPa can be also found on several websites including:

http://www.weatheroffice.gc.ca/data/analysis/sai_50.gif

http://www.hbc.co.jp/pro-weather/AUXN50-jpg.html

Some useful guidelines for interpreting weather charts can be found on the website of the Center for Ocean-Land-Atmosphere Studies, COLA (http://wxmaps.org/pix/fcstkey.html).

The geopotential height forecasts from NCEP and ECMWF for D+4 at 500 hPa appear in the upper panels of Fig. 4.2.3 and Fig. 4.2.4. They are verified by NCEP analysis at 12 UTC on 13 September 2008. The forecasts exhibit common features in the large-scale patterns, but some differences in local features. Both forecasts generally agree on the phase of midlatitude troughs, and Hurricane Ike with some discrepancy on the short trough off the coast of southern California. The forecast difference comes from the difference in computer models and the initial states they start with. The degree of uncertainty in model forecasts is visualized in the upper panel of Fig. 4.2.4 using a spaghetti plot of geopotential contours from the ensemble of model forecasts with varying initial states. The spread of model forecasts on the position and intensity of the upper-level trough near California is partly associated with the poor simulation of the landfalling Hurricane Ike near Texas. More details on the ensemble forecast will be discussed in Chapter 6.

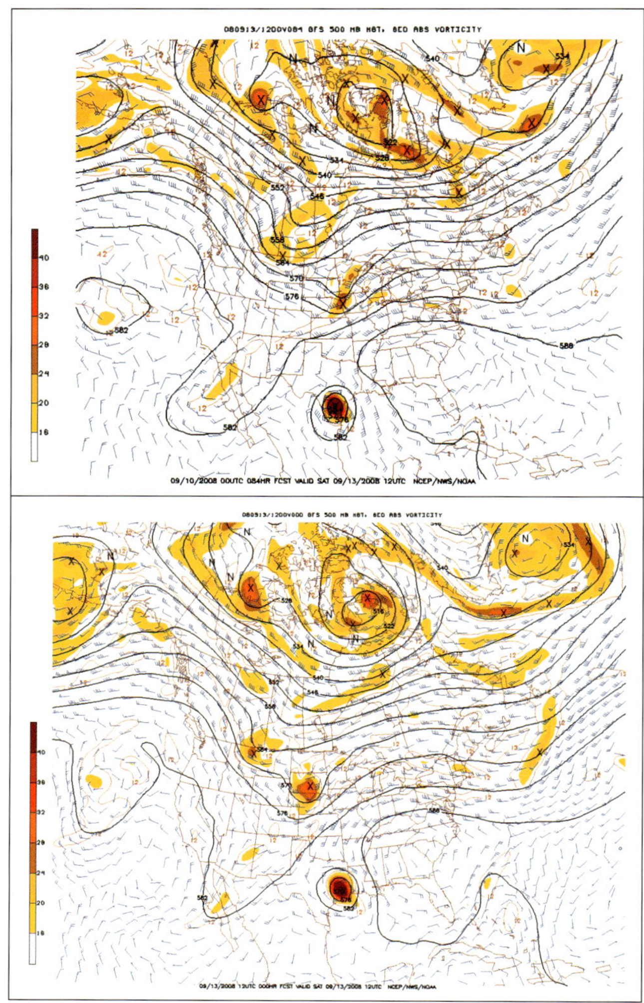

Fig. 4.2.3. Computer model forecast chart for 84 hr projection from NCEP (upper) verified with analysis (lower) valid at 12 UTC on 13 September 2008. The contour lines are the geopotential at 500 hPa, and the colors represent the relative vorticity. The wind feathers are also presented in the analysis and forecast charts. The images are available from http://mag.ncep.noaa.gov/NCOMAGWEB/appcontroller. Source: U.S. NOAA.

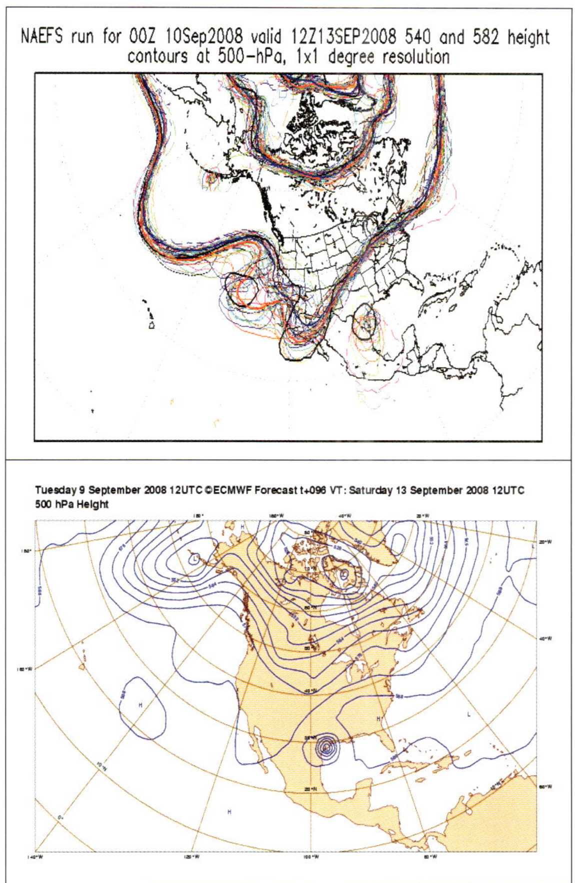

Fig. 4.2.4. Computer model forecast chart for 84 hr projection from NCEP (upper) and 96 hr projection from ECMWF (lower) respectively valid at 12 UTC on 13 September 2008. The contour lines are the geopotential at 500 hPa. In the upper panel, the contours of 84hr ensemble predictions from NCEP, U.S. NOAA are presented in the form of a spaghetti diagram. The images are available from http://www.emc.ncep.noaa.gov/gmb/ens/, and http://www.ecmwf.int/products/forecasts/d/charts/medium/deterministic/.

Similar charts for various regions can be found on numerous websites, some of which are listed in Table 4.2.1. It is noted that many commercial sites link to analysis and forecast charts of the national meteorological centers. These are not mentioned in the table for brevity.

Table 4.2.1. Examples of websites that provide detailed analysis and forecast charts for mean sea level pressure (MSLP), geopotential height (z) at 500 hPa, streamline, and thickness (z) between 1000-500 hPa. Some websites also provide other field variables such as temperature and humidity. The 00 hour in the type refers to analysis.

Region	Type	Addresses (http://)
N. Atlantic	MSLP and 1000-500 z for 00 to +120 hours	www.weather.org.uk/charts/UKCpf000.gif; UKCpf024.gif; UKCpf048.gif; UKCpf072.gif; UKCpf096.gif; UKCpf120.gif
S. Pacific	MSLP and z for 00 to 144 hours	www.bom.gov.au/australia/charts/viewer/ www.bom.gov.au/difacs/IDX0032.gif
S.Indian Ocean	MSLP for 00 hour	www.bom.gov.au/difacs/IDX0033.gif
Tropics	Streamline	www.bom.gov.au/cgi-bin/nmoc/latest_D.pl?IDCODE=IDX0014
Europe	MSLP for 00 and +36 hours	www.metoffice.gov.uk/weather/uk/surface_pressure.html www.met.fu-berlin.de/de/wetter/maps/anabwkna.gif www.met.fu-berlin.de/de/wetter/maps/emtbkna.gif 131.54.120.150/index.cfm?section=dspImage&image=21OWS_EUROPE_FITL_PROG-BRIEFING_12.gif
Global	MSLP and z for +72 to +240 hours	www.ecmwf.int/products/forecasts/d/charts/medium/deterministic/ www.fnmoc.navy.mil/wxmap_cgi/index.html www.opc.ncep.noaa.gov/UA/entire_UA.gif www.weatheroffice.gc.ca/model_forecast/global_e.html www.bom.gov.au/nmoc/NWP.shtml
Asia	MSLP & 500 z for +00 to 240 hours	web.kma.go.kr/eng/weather/images/nwp_intro.jsp www.jma.go.jp/en/g3/
N. America	MSLP	www.weatheroffice.gc.ca/data/analysis/jac00_100.gif

4.3 Surface charts

The weather is closely related to the pressure patterns near the surface of the earth. The contours of mean sea level pressure are drawn in surface charts. The surface pressure over mountains is hypothetically extrapolated from adjacent model levels down to the mean sea level to examine atmospheric condition on an equal footing. The validity of pressure and temperature charts over high mountain regions heavily depends on the effectiveness of the assumed lapse rate and stability, and topography used for the derivation of the mean sea level pressure.

Low-pressure areas (L) are in most cases associated with the sensible weather such as rain, snow, strong wind, etc. High-pressure areas (H) correspond to fair weather, but are sometimes accompanied by fog or stagnant pollution. The wind blows counter-clockwise around surface lows and clockwise around surface highs in the northern hemisphere, as illustrated in Fig. 4.3.1. The reverse applies in the southern hemisphere. Wind direction depends on the position of the observer relative to the center of moving low or high pressure areas. For instance, an observer sitting in the one o'clock position from the center of the low in the northern hemisphere feels south easterlies, while one in the 7 o'clock position of the same low observes north westerlies. Wind speed is directly proportional to the packing of the isobars, in other words, intensity of lows or highs. The pressure gradient force conditions being equal, the wind around highs is normally stronger than around lows.

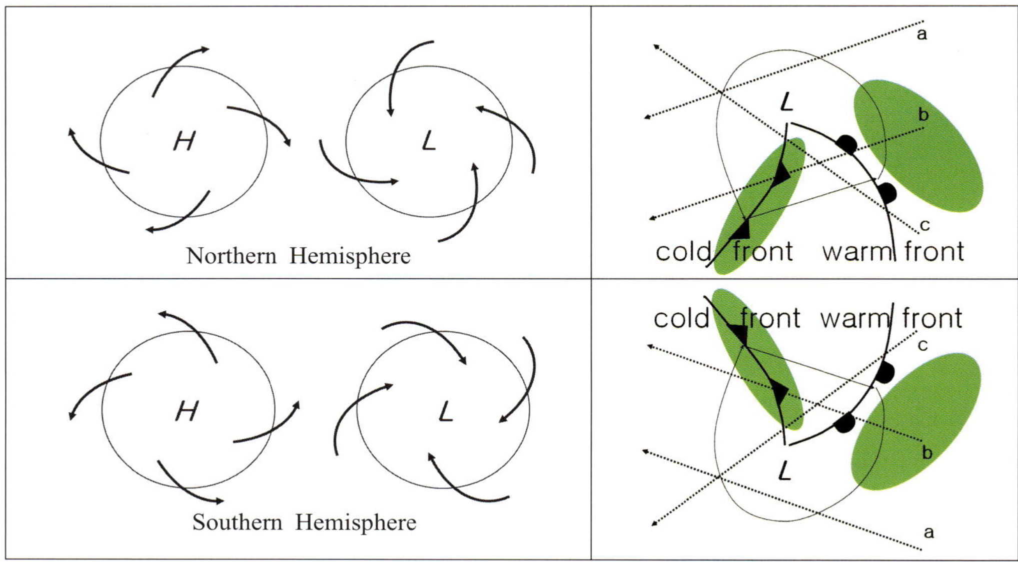

Fig. 4.3.1. Schematic diagram of wind circulation (solid arrow) around high and low pressure systems and fronts in the northern hemisphere (top), and the southern hemisphere (bottom). The shaded area indicates the preferred location for precipitation associated with fronts. The dotted arrow indicates the relative motion of the observer at locations a, b, and c from the moving low. An observer (a) northeast of the low in the northern hemisphere feels southeasterlies first, then sees them changing to northeasterlies later. An observer (c) southeast of the low observes southeasterlies, southwesterlies, and finally northwesterlies. By the same token, an observer (a) southeast of the low in the southern hemisphere feels northeasterlies first, and sees them changing to southeasterlies later. An observer (c) northeast of the low observes northeasterlies, northwesterlies, and finally southwesterlies.

It is noted that the surface low in an actual chart never appears in the form of a perfect circle. It is often elongated in a certain direction. At times it is not closed, and overshadowed by the dominant high. In any case it is always important to check where warm and humid air comes from, as its source has better chance of sensible weather than others.

Front models

A front is represented by a thick continuous line, as shown in Fig. 4.3.2. A warm front is marked using a red line with semicircles pointing toward the direction the front is moving. A cold front is marked using a blue line with triangles pointing toward the direction of the front. An occluded front is marked using a purple line with alternating triangles and semicircles pointing toward the direction of the front. A stationary front is marked using alternating blue triangles and red semicircles on opposite sides. Unfilled triangles or semicircles indicate that the front is aloft or not on the ground.

	Warm front		Occluded front
	Cold front		Stationary front

Fig. 4.3.2. Map symbols for fronts reproduced from Appendix II-4 in the manual on the global data-processing system of WMO.

A conceptual model of fronts is a useful starting point in forecasting weather accompanied by an extratropical cyclone in the midlatitude, as illustrated in Fig. 4.3.1. The southerly wind brings warm air over the forward flank of the low, while the northerly wind pulls down the cold air on the rear end of the low in the northern hemisphere. The boundary between the two air masses first forms a wedge and then develops a front-like structure. Frontal type and associated weather can be determined by the character of advancing air mass. Warm air, overrunning the cold air mass, moves in line with the warm front. The cold front progresses in the direction in which the cold air penetrates below the warm air mass. Sometimes the two air masses confront each other without movement in either direction and form a stationary front. Most weather disturbances occur near the fronts. Gentle clouds and mild precipitation occur around warm fronts; violent squalls with gusty winds are frequently observed near cold fronts. In the warm sector surrounded by the two fronts, severe thunderstorms often form in summer time.

Weather systems in the midlatitudes mostly move from the west to the east. The front lines that are inclined at a wide angle to the latitude circle tend to move eastward faster with westerlies. On the other hand, most stationary fronts are at a narrow angle with respect to the latitude so that they are hardly carried up by westerlies. Stationary fronts are often associated with flooding as precipitation persists for a longer time in any location. A monsoon trough is a favorable location for a stationary front in the northern summer.

Wild fires often start with lightning thunderstorms accompanying cold fronts. If the upper-level ridge and surface high pressure persist in an area, the atmospheric condition of high temperature and low humidity increases the risk of fires to critical levels. The approaching upper-level trough upstream will increase the instability of the atmosphere, and lightning accompanied by thunderstorms ignites fires in dry areas. Gusty winds can cause fires to spread faster, especially a few hours before the frontal passage. Cooler air and high humidity from rain or snow usually help reduce the risks of fire behind the front.

Changing weather affects physiological and pathological conditions. Approaching high pressure systems are generally associated with quiescent weather phases, and do not affect the human body very much if the latter is equipped with proper clothes and sun protection. However, local pollutants mix with the air in the ending phase of a high pressure system, if it persists for many days supporting stable conditions. Fog often combines with smoke and may add stress on the human respiratory system. Heat stress increases with the passage of warm fronts as abundant moisture is supplied from the subtropics. On the other hand, the wind chill index goes up rapidly at the rear of a cold front where strong polar winds blow into the low center.

The passage of cold fronts brings turbulent and violent weather that affect our health (Landsberg, 1986). In some cases, this abrupt change may cause heart attacks, ulcers, migraines, spasmodic diseases and others. Gusty winds generated by thunderstorms or cold fronts stir up sand, pollen, spores and fungi and cause hay fever, asthma, and other respiratory diseases. Some people suffer from storm phobia (Kleinknecht, 2002) characterized by hyperalertness and vigilance in responding to thunders and gusty winds with the passage of low-pressure systems.

In the example of Fig. 4.3.3, two cold fronts are found over the west Canadian border and mid-western plains. Three low-pressure systems appear on the chart. The one in Texas is nothing but a landfall of Hurricane Ike. The other one, centered over Iowa, forms a stationary front extending to the east that is responsible for the rain band in Fig. 1.2.2. The last one centered on the Canadian border of North Dakota is connected with a trailing cold front that collocates with the separate rain clouds on North Dakota and Montana. The rain bands in Fig. 1.2.2 correspond well to the frontal zones or to the surface low. The consistency between the contours and neighboring observations in the lower panel of the figure, satisfying the gradient wind relationship, partially corroborates the analysis. Certain types of observations such as satellite radiance are not visualized directly in the analysis chart, even though they are used for data assimilation. Those contours, unless specified otherwise, are drawn first by the computer with numerical algorithms using the physical relationship among variables, and are then updated further by subjective analysis. Additional details on computer analysis will be discussed in the next section.

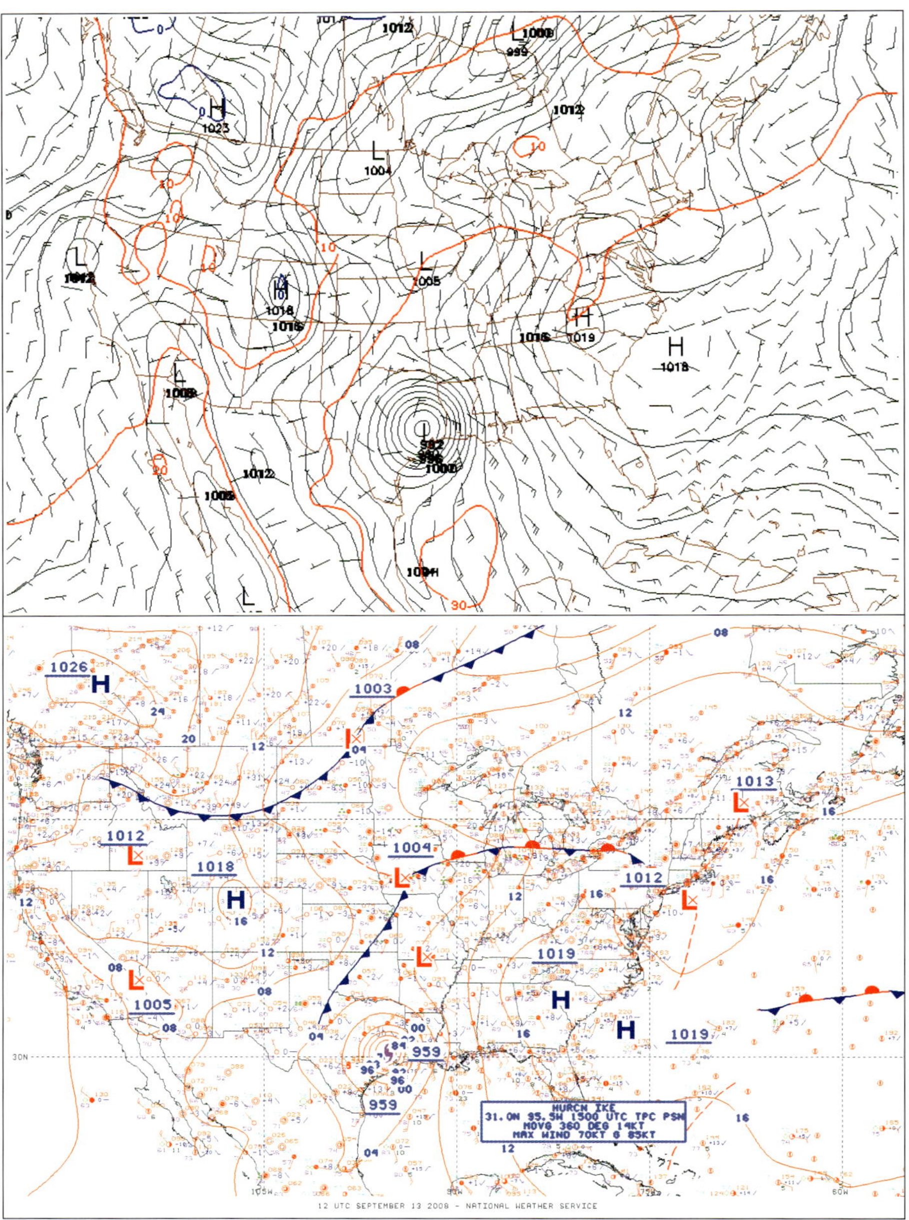

Fig. 4.3.3. Two types of weather charts for 12 UTC on 13 September 2008; (Upper) computer generated analysis for mean sea level pressure with contour intervals of 2 hPa, wind (10 m above ground), and temperature (2 m above ground) with contour intervals of 10 degrees Celsius, and (Lower) surface analysis chart with manually updated fronts. The images are available from http://mag.ncep.noaa.gov/NCOMAGWEB/appcontroller and http://www.opc.ncep.noaa.gov/UA/USA.gif respectively. Source: U.S. NOAA.

Weather symbols

Observations at each ground station are represented on surface charts in graphical format following WMO guidelines, as illustrated in Fig. 4.3.4. The center circle represents the station. The wind feather points the direction where the wind comes from. Barbs express the wind speed. Each half-barb represents 5 knots (9 km/hr), and each full-barb 10 knots (19 km/hr). Each flag or filled triangle corresponds to 50 knots (93 km/hr). Cloud cover is represented inside the circle to which the wind feather points. It ranges from cloudless (0/10) to overcast (10/10), and is proportional to the filled area inside the circle. The station symbol includes numbers surrounding the cloud circle. The barometric pressure is shown at the upper right corner. The number on its right indicates the pressure change. The temperature and dew point are indicated on the left with visibility between them.

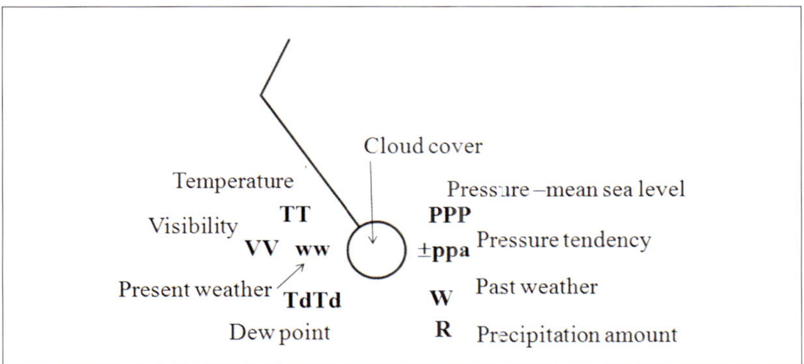

Fig. 4.3.4. Station symbols.

A symbol for the current weather is also shown at the left. Rain and snow are symbolized by '·' and '*' respectively. Fog is denoted by '≡'. More symbols are shown in Table 4.3.3. The full list of symbols on weather can be found in the Appendix II-4 of the manual on the global data-processing system of WMO.

Upper and lower cloud symbols are shown above and below the cloud circle. A number to the right of the lower cloud symbol denotes the fractional sky cover in coded form. The weather in the recent past is on the lower right.

Table 4.3.3. Frequently used weather symbols in surface weather charts reproduced from the Appendix II-4 in the manual on the global data-processing system of WMO.

•	Rain	⚡︎	Thunderstorm
,	Drizzle	≡	Fog
✱	Snow	S→	Dust or sand storm
▽	Showers	∞	Haze
△	Hail	⌐~	Smoke

Figure 4.3.5 illustrates station symbols from observation sites in the United States. It is overcast with continuous light rain at Burlington, which is located in front of a cold front travelling over Kansas from the northwest. The wind is 5 knots from the southwest. The temperature is 74F (23.3℃), and dew point temperature is 71F (21.7℃). The barometer reads 1006.0 hPa. Behind the cold front, the weather is clear at Hill City with a relatively high pressure of 1008.7 hpa. The barometer has fallen 0.3 hPa during the last three hours but has since started rising. The wind is 5 knots from the northwest. The temperature is 59F (1 5℃) and dew point temperature is 54F (12.2℃), which indicate that the cold and dry air mass from the northwest has already arrived at Hill City. On the other hand, Chicago, under the influence of a stationary front, records heavy rain since the last report with temperature at 69F (20.6℃) and dew point at 68F (20℃). Surface chart weather symbols can be found at numerous websites including:

http://www.weatheroffice.gc.ca/analysis/index_e.html

http://www.opc.ncep.noaa.gov/UA/USA.gif

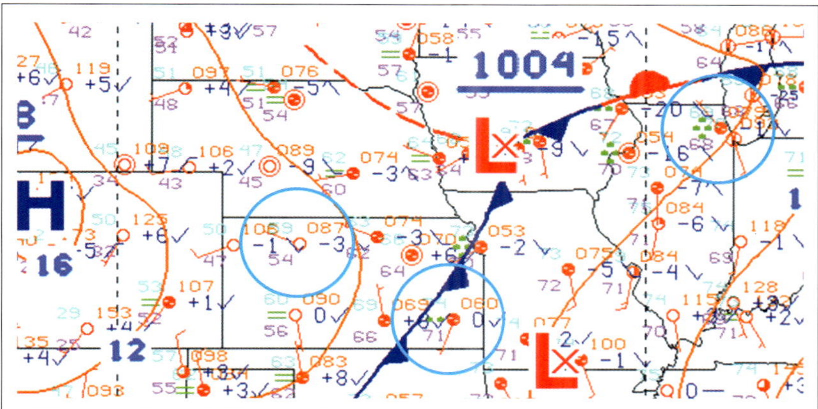

Fig. 4.3.5. Enlarged weather map from the lower panel of Fig. 4.3.3. Surface chart weather symbols near Burlington, Chicago, Hill City are highlighted in blue circle.

Many centers provide forecast precipitation on MSLP charts. Specifically, the US Navy website provides regional forecast charts for various theaters around the world. An example from NCEP is shown in Fig. 4.3.6. One should take note of the period of rainfall accumulation in those charts, which is normally indicated on the side or bottom of the chart. Some use 6-hr or 12-hr, while others use 24-hr. The rainfall pattern includes small-scale noises comparable to the grid intervals used in the numerical calculation. They are likely to be erroneous. It would be safe to concentrate on the large-scale patterns and trends when reading forecast rainfall charts.

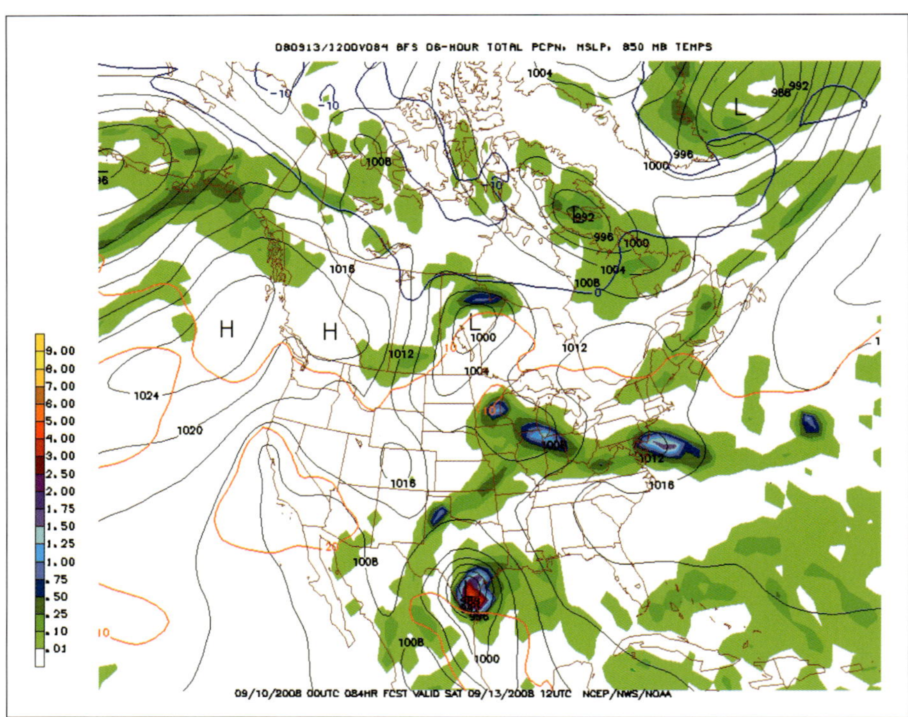

Fig. 4.3.6. Example of a 96-hr prognostic chart valid at 12 UTC on 13 September 2008 for mean sea level pressure (black line) and temperature at 850 hPa (red line), and 6-hr accumulated precipitation (color filled) from NCEP computer models. The image is available from http://mag.ncep.noaa.gov/NCOMAGWEB/appcontroller Source: U.S. NOAA.

The precipitation forecast in Fig. 4.3.6 generally matches observations in Fig. 1.2.2 in areas such as Texas affected by Hurricane Ike and northern plains affected by fronts. However, the rainfall over North Dakoda is displaced southeastward in the model. The large-scale pattern of precipitation can be predicted as many as 4 days ahead. The precipitation field in 6 hr or 12 hr prognostic charts is much closer to the real atmosphere, and is often used for the frontal analysis as shown in Fig. 4.3.3.

The conceptual models of fronts in Fig. 4.3.1 and vertical coupling in Fig. 4.2.1 are not applicable to tropical weather systems. Squalls and organized convection easily develop in the tropics under the conditionally unstable atmosphere with weak synoptic forcing. The low-level pressure or wind patterns by and large control the local weather in the tropics, and the upper-level signals are not obvious in many cases. Tropical cyclones are not an exception. The circular wind near the eye wall clouds of a tropical cyclone is strongest near the surface and decreases with height. This is primarily associated with the unique thermal wind relationship in the warm core cyclone under the unstable air mass; i.e., the warm air occupies the center and the moderately cool air is outside.

4.4 Computer analysis

The forecast charts discussed in the previous sections are derived from computer analysis charts. Understanding computer analysis charts and their characteristics help interpret current weather conditions and forecast charts. Computer analysis is not made from scratch. New observations exert an incremental effect on a priori information. The computer analysis is updated by observation increments on top of the background field. In most cases very short range model forecasts are used as the background field for the computer analysis, assuming that the model forecasts satisfy the complex nonlinear balance given by the governing principles of primitive equations.

The observation increment is defined in Fig. 4.4.1 as the difference between the observed and the background field on the model grids. The observation increment is weighted by the ratio of background error over the total error, i.e., the sum of background error and observation error. More credit is given to the observation where the background error is large. The background error covariance also determines the relationship between observation sites. Observations in data-sparse areas are weighted more than those in densely populated areas. The fundamental balance constraint, including geostrophic and hydrostatic approximation valid for large-scale motions, is implicitly incorporated in the background error covariance. A change in one variable leads to a corresponding change in the other variables.

Four-dimensional variational data assimilation (4dVar), a state-of-the-art data analysis scheme considers dynamic balance among observations taken at different times and locations.

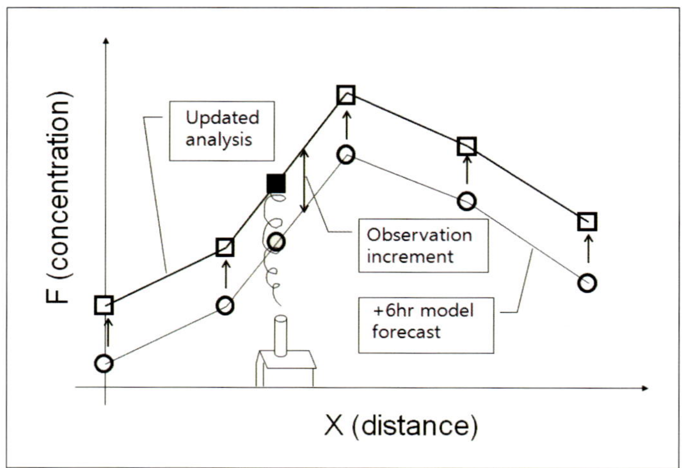

Fig. 4.4.1. Schematic for the analysis of dust concentration F, as an example, utilizing latest observation (filled square) and background field (circles) at grid points in x-direction. 6hr forecast from the previous model run is used for the background field. The observation increment is evaluated from the measurement at observation sites (filled square) minus the background field value. The observation increment is then extrapolated to neighboring grid points (open squares) with the statistical weights derived from the error covariance matrix of the background field normalized by observation error.

The quality of the computer analysis has to be carefully monitored where the stationary background error covariance is violated or the background field poorly represents the underlying physics. First, the background error used in the data assimilation is obtained from past samples of model simulations. Transition states under changing weather regimes are outliers in the sample space, and their background error characteristic would not be well reflected in the averaged statistics. Ensemble Kalman filter data assimilation may help alleviate the sampling problem, taking into account in situ background error with flow dependency.

Second, numerical models are very sensitive to the initial atmospheric conditions, and the dynamic chaos inherent in the model atmosphere amplifies the initial errors beyond the predictability limit. Fine-scale systems, supported by ageostrophic forcing, undergo a complex physical process. They often evolve with the cycle of rapid development and decay, which results in large errors in the background field. Even high-resolution models using non-hydrostatic equations have certain limitations in resolving detailed mesoscale features near the surface and mountainous regions with turbulent atmospheric motions.

Humidity analysis needs particular attention. Humidity is poorly observed, and moist physics interacting with clouds and the planetary boundary layer (PBL) is not well understood. Simplification of the model used in variational analysis further limit the resolution of the analysis field. For example, the forecast moisture field is systematically distorted due to ill-posed surface boundary conditions originating from the unrealistic forcing of precipitation and clouds. These defects are transferred to the background field, and contaminate the analysis after many cycles of data assimilation. Users must check the validity of computer-generated moisture fields in light of ground measurements and other remote sensing observations available.

The quality of contour analysis depends on the computer model, skill of the analyst, and the observations used. Human intervention is particularly instrumental to improving the quality of analysis near the surface and fast moving systems. Time restrictions, however, make it difficult for human analysts to do quantitative comparisons of remote observations and synoptic analyses. Over the ocean and mountainous regions where synoptic observations are sparse, users are advised not to correct the computer-generated analysis simply based on conventional observations. In addition, users are requested to give a balanced credit to both the model background and the latest observations to optimally capitalize on short-term model predictions and associated physical principles.

5. Quality of Weather Information

5.1 Data mining

The information we seek exists somewhere on Earth. We might be looking for introductory notes on meteorological terms, research papers, teaching materials, interpretation of meteorological data, or local observations. A considerable amount of data is sitting unused on the Internet, waiting for potential users. With enough time, data mining can locate almost any kind of information on the Internet. Language barrier may impede access to online information in some cases. However, no such obstacle exists for weather charts. One only has to learn how to interpret the contours and symbols on a weather chart. As far as meteorological data is concerned, one cannot claim ownership because someone else may upload equivalent information independently on the Internet. The real owners are those who know where to find the right information and how to use it. Application, not possession, determines the value of meteorological information.

It doesn't take long to realize that similar meteorological information is out there once we try data mining on the Internet. Some of the information merely comes from multiple links that connect to the same data source, while some versions are the products of different ways of processing of the same input. Weather forecasts for New York can be found not only on the U.S. NWS homepage, but also on websites of newspapers or portals in any other country. Surface charts or 500 hPa charts can be found on several websites, as illustrated in Table 5.1.1. Seasonal mean anomaly charts at 500 hPa are also available at many global modeling centers as shown in Table 3.2.1. The weather information from each website may differ in terms of content, style, and presentation. Confusion arises when one hears different voices on the same subject. It is stressful to determine which one is good or bad, and to evaluate which one is better. A person may be ambitious enough to synthesize all of the available information by taking into account the credibility of the source, if sufficient time is allowed for the analysis. However, it is often the case that the degree of reliability or uncertainties of the information is neither known, nor accessible without permission.

Table 5.1.1. Some websites providing access to computer model products for the regional domain. For more information, visit the WMO homepage at http://www.wmo.int/pages/prog/www/DPS/gdps.html

Meteorological Center	URL (http://)
BoM (Australia)	www.bom.gov.au/australia/charts/
CMC (Canada)	www.weatheroffice.gc.ca/model_forecast/global_e.html
ECMWF	www.ecmwf.int/products/forecasts/d/charts/medium/deterministic/
INMET (Brazil)	www.inmet.gov.br/html/prev_tempo.php?lnk=../prev_clima_tempo/regioes_p.html
JMA (Japan)	www.jma.go.jp/en/g3/index.html
KMA (Korea)	www.kma.go.kr/ema/ema03/ra2_eng_index.html
NCEP (USA)	mag.ncep.noaa.gov/NCOMAGWEB/appcontroller
SAWS (S. Africa)	www.weathersa.co.za

Open data available online will continue to increase, as data providers compete with each other to attract users. The burden is on the users to choose the best source in the sea of information on the Internet. The weather business is a war against time. The value of meteorological information plummets as time goes by. Climatological data services are an exception. Like news on TV, a weather forecast loses most of its value past the valid forecast time. The battle is between the time spent on analysis and the volume of data to be digested. More data used for analysis may lead to more reliable information. But the opportunity cost increases as well due to the extra time spent for analysis. Promptly processed information allows one to respond ahead of the potential adverse weather event. But the uncertainty of the information increases as well due to the shortage of supporting data.

The flood of data is challenging to professional meteorologists and laypersons alike. The number of basic charts used to investigate short-range forecasts easily exceeds hundreds. One should examine analysis charts at various pressure levels such as 925, 850, 700, 500, 300, 200 hPa, and the surface. The temperature, humidity and wind should be examined along with geopotential height. Many diagnostic variables supplement the analysis, such as vorticity, vertical motion field, and vertical stability, etc. Visit the websites of the U.S. Navy Operational Global Atmospheric Prediction System (NOGAPS), NCEP, and KMA for instance, and you will soon realize that hundreds of charts are waiting for your click.

Extending data sources to the private sector, the problem of data redundancy becomes even worse. Many research institutes and university departments upload the latest outcomes of research projects or teaching materials on the Internet. Some institutes regularly produce forecast charts near real-time and provide them online. For instance, NCAR runs a numerical model testbed for different climate zones and presents the model outputs on its website. The International Research Institute for global change studies (IRI) runs a climate model and provides simulation results on its website along with other products from other centers. The Typhoon Research Center at Kongju National University uploads on the web track and intensity forecasts from typhoon prediction models. Many commercial firms also provide various observations and forecasts in text or graphic format, some of which are aimed at promoting product sales and maintaining a high profile as a leading science organization. Generally speaking, many of the forecasts and other information from the private sector are very up-to-date and often creative, reflecting the latest science and technology. However, the reliability of the data is not proven for application in many cases and users should be cautious in using those data. Another concern is the stability of the data service. Information updates may be abruptly interrupted at any time without prior notice. Some of them may be transformed to commercial services once their utility has been proven, which may prove an unpleasant surprise to the Internet users who had previously accessed the data free of charge.

5.2 Forecast verification

Meteorological contents in the form of charts, figures, and tables are becoming more attractive, with colorful designs as multimedia and graphic technology advances. One must note, however, that attractive presentation does not necessarily mean reliable information. The quality of the information is not guaranteed unless supported with evidence of validity. It is not surprising to learn that most of the freely accessible information generally shares the common problem of proving credibility, except for the government-owned or highly respected organizations. Some national meteorological centers provide verification scores along with their forecasts on the web. Credibility is a fundamental attribute of any information and is the price to be paid for its purchase. Users who rely on data freely available on the Internet must develop their own skills to distinguish reliable from unreliable data and interpret them in the context of decision-making situations.

Insights on the forecasting process help one understand factors affecting the credibility of a given forecast. A weather forecast, as discussed in-depth in Chapters 3 and 4, is the end product of a scientific process that consists of observation, analysis, numerical prediction, interpretation, and dissemination, whose complex nature is revealed when a forecast's reliability is evaluated. The credibility of a weather forecast depends on the quality of the by-products at each stage of the forecasting process. The research infrastructure and software engineering, such as the management of the forecast office, coordination of nationwide research, and efficiency of mechanisms for transferring academic findings to operation, may influence the quality of a forecast.

All those components add complexity to the evaluation of credibility of meteorological information. Recently the Commission for Basic Systems (CBS) of WMO considered introducing a certification procedure such as ISO 2000 for forecast services. Until the new certification scheme is in place, the users must themselves evaluate the quality of weather forecasts and the forecast charts that they are based on.

The reputation of the weather service bureau may be used as an indicator in estimating the credibility of a forecast if no other information is available. Weather forecasting as a discipline is considered a multidisciplinary science that includes weather analysis and forecasting, remote sensing by radar and satellites, interactive communication using the Internet and multimedia, software development for office automation and production, and supercomputer simulation of geo-fluids. The quality of weather service reflects in some sense the status of science and technology and infrastructure at the national level. From a macroscopic viewpoint, one may reasonably say that weather forecast and products from well-known organizations or government-supported institutes in advanced countries are more credible than others, other conditions being equal.

The easiest way to assess the credibility of a forecast chart is to see what the professional forecasters use in daily operations. Many national meteorological centers use an ensemble of forecast charts from their own and other national meteorological centers as shown in Tables 3.2.1 and 5.1.1. Based on daily experience and regular assessment of forecast errors, forecasters build a knowledge base to choose the best model output. Forecast accuracy also depends on the variables. Preference for forecast variables is different from forecaster to forecaster. Forecast charts of geopotential and vorticity at 500 hPa are commonly used for short and medium range forecasts. Temperature forecast charts at 850 hPa are also widely used. Precipitation forecast charts are extensively used along with mean sea level pressure contours, but their accuracy is much lower than others.

It is not easy to compare the quality of weather forecasts from different sources under the same criteria. How weather forecast quality is measured differs across forecast producing centers. One way of evaluating the quality of weather forecasts is to check model forecast errors. Even though forecasters' knowledge and expertise are a substantial part of the decision process, their role is limited in most cases to partial compensation of flaws in the model output. In other words, weather forecast accuracy pretty much depends on the performance of the computer model. It is noted, however, that the verification scores of weather forecasts are not directly linked to specific computer model forecasts, as the official weather forecast from any national meteorological service is in general based on a multitude of computer model forecasts from various centers. Figure 5.2.1 demonstrates that forecasters contribute about 10~25% of the variance of observed weather explained by the short-term forecasts, and the rest results from the computer models. The lower panel of the same figure indicates that the official precipitation forecasts have gradually improved in parallel with the model precipitation forecasts. The forecasters have better skill than the computer model, particularly on the very short range up to 6 to 12 hours in advance. Further discussion will be given in Chapter 9.

Fortunately, major meteorological centers regularly evaluate the performance of their computer models in accordance with the standard verification procedures of WMO. Some information on the accuracy of computer model products can be accessed on the WMO website for D+1 to D+7:

http://www.wmo.int/pages/prog/www/DPFS/ProgressReports/index.html.

Similar quality information is produced for the long-range prediction model forecasts from each climate modeling center. For contact points and detailed verification procedures for long-range forecasts, visit the WMO website at:

http://www.wmo.int/pages/prog/www/DPS/gdps.html.

The standard verification indices for short-range forecasts differ from those for long-range forecasts. Some common indices are explained below.

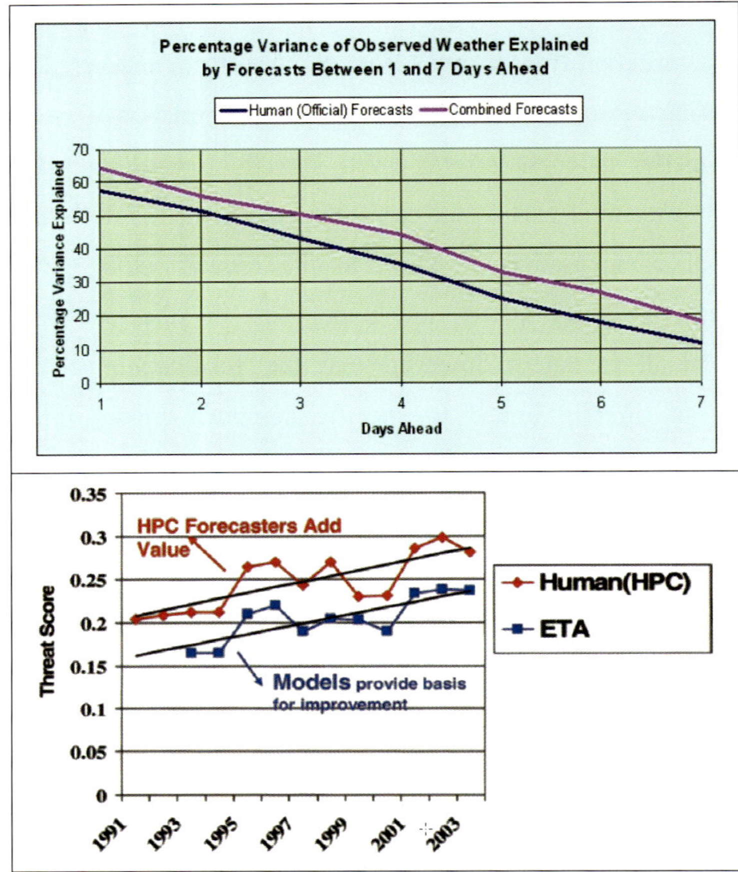

Figure 5.2.1. (Upper) Overall percentage variance of observed weather explained for each forecast lead time explained by forecasts between 1 and 7 days ahead. The weather elements of rainfall amount, sensible weather, minimum temperature, and maximum temperature are taken together for the evaluation. The combined forecasts displayed an improvement in the accuracy of forecasts for each forecast lead time (Stern, 2005). (Lower) Comparison of threat scores for D+1 precipitation between forecaster and the Eta model from the NCEP/HPC. Human forecasters add about 20% of value to numerical model forecasts (Harper et al., 2007). The threat score is defined in (5.2.3).

Commonly used indices for measuring the accuracy of 2-dimensional forecast charts include the root mean square error (RMSE) and the anomaly correlation (AC). The RMSE and AC for the geopotential height at 500 hPa are globally used for the comparison of forecast charts. RMSE is the squared root of the average of individual squared differences between the predicted and the observed at selected points. The RMSE tells us the degree of departure of the prognostic field against the analysis or the observed. The RMSE is often used to evaluate the error magnitude in a forecast of continuous variables such as temperature, winds, and geopotential heights.

$$RMSE = \sqrt{\frac{1}{N}\sum_{i=1}^{N}\left(F_i - O_i\right)^2} \tag{5.2.1}$$

Here N is the number of all points in the domain. F_i is a predicted value at i's point. O_i is the corresponding value of the observed at the valid time. RMSE is always greater than zero. The more the RMSE increases, the more the corresponding prognostic chart deviates from reality. When the RMSE becomes zero, the prognostic chart perfectly matches with the observation. When a model predicts high and low completely out of phase to the observed, its RMSE marks higher than the null forecast. The RMSE depends on the natural variability of the flow. The same model yields different RMSE values from season to season. The RMSE of the computer forecast chart is generally low in summer when wave activities are weak. The minimum temperature forecast is less accurate in winter, specifically for the eastern coasts of major continents. The maximum temperature forecast shows similar deficiency during spring in the mid-latitude zone. One should understand such characteristics when interpreting the RMSE as a measure of model performance (Persson, 2001). The RMSEs of model forecasts from various centers are presented in Fig. 5.2.2 for reference. The RMSE can also be used for the measurement of position error in tropical cyclone forecasts. In that case the formula (5.2.1) refers to the summation over the forecast samples.

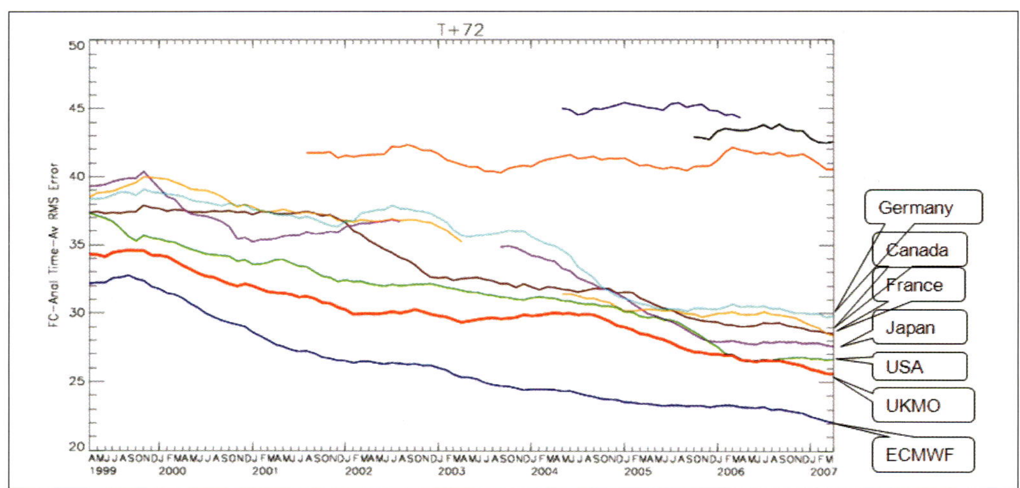

Fig. 5.2.2. Trend of root mean square error (RMSE) of D+3 computer forecast from 10 operational centers in units of m for geopotential height at 500 hPa in the Northern Hemisphere during 1999~2007. Source: ECMWF.

The AC is a measure of the similarity between the predicted and the observed anomalies in terms of pattern or shape of contours on a 2-dimensional surface. Here the anomaly means the departure from the climatological mean. The average over 10 years or more is typically used for climatology.

$$AC = \frac{\sum_{i=1}^{N}(F_i - C_i)(O_i - C_i)}{\sqrt{\sum_{i=1}^{N}(F_i - C_i)^2}\sqrt{\sum_{i=1}^{N}(O_i - C_i)^2}} \tag{5.2.2}$$

where C_i is the climatological value at the i's point.

An AC equaling one indicates a perfect forecast where every trough and ridge line in the prognostic height forecast chart matches with those in the verifying analysis. An AC below 0.6 is in general regarded as poor quality for short and medium-range forecasts. The AC is not as sensitive to the natural variability of the flow as the RMSE. However, the AC tends to yield higher scores in the low zonal index, or north-south flow pattern, as it reflects shape match. When a model fails to predict the correct phase in the zonal flow pattern, the AC can easily plunge down to lower values than in the meridional flow pattern (Persson, 2001). The AC of the computer forecast chart is generally high in winter when the upper level jet and associated wave activities are stronger.

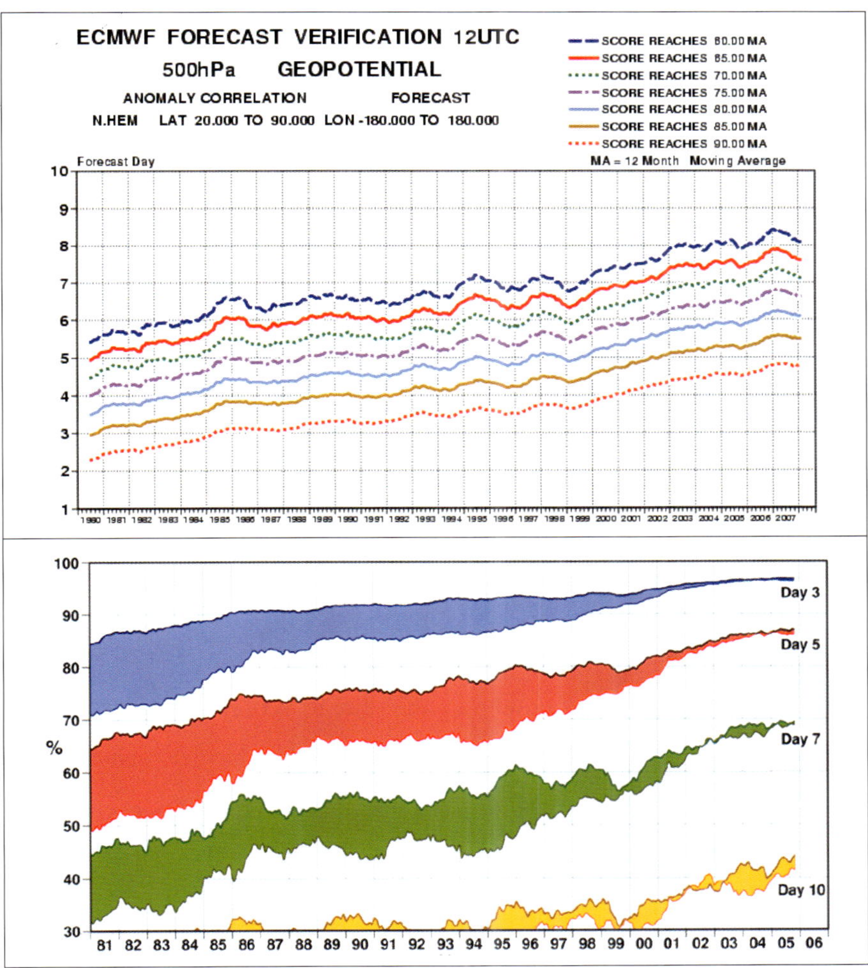

Fig. 5.2.3. Anomaly correlation in terms of lead-time exceeding various threshold levels ranging from 60 to 90% in the northern hemisphere (upper), and each hemisphere for geopotential height forecast (lower) at 500 hPa from ECMWF. The upper and lower bounds in the lower panel refer to the Northern and Southern Hemispheres. The image in the upper panel is available from http://www.ecmwf.int/products/forecasts/d/charts/medium/verification/timeseries/z_score_monthly_me

an_runmean!North%20hemis.!12!pop!od!oper!w_scores_mean!latest!/. The lower panel is taken from a presentation slide by Dr. A. Simmons, ECMWF.

A forecast chart has some value as long as its AC exceeds at least 0.6. The AC decreases as forecast lead time increases. Predictability is often defined in terms of the forecast lead time at which the AC first drops down below 0.6. Figure 5.2.3 presents the AC in terms of lead time exceeding various thresholds for geopotential height forecast at 500 hPa from state-of-the-art computer model forecasts. ECMWF predictability is about 8 days, showing a gradual improvement of skill score since the early 80s. As the science and technology of numerical weather prediction advances, the AC of the D+8 forecast nowadays corresponds to what the model scored for D+5 forecasts 25 years ago. The southern hemisphere has been catching up with the northern hemisphere in recent years. The progress is largely due to advanced data assimilation of satellite radiance observations. Forecast accuracy over the tropics is not presented in the figure, but is reported to be inferior to compared with mid-latitude forecasts. The difference is mainly due to the sparse observation network and poor understanding of physics over the tropical atmosphere.

The reliability of precipitation forecasts is measured in a different way. The critical success index (CSI), known as the threat score, is widely used in the scientific community. It is the ratio of hit cases over the test samples excluding the correct negatives, i.e., cases with no rain forecasts and no rain observed.

$$CSI = \frac{H}{H + M + F} \tag{5.2.3}$$

Sometimes the correct negative is considered in the measure of accuracy (ACC), i.e.,

$$ACC = \frac{H + C}{H + M + F + C} \tag{5.2.4}$$

Where H and F are numbers of hits and false alarms respectively for a given forecast of an event occurring. In Table 5.2.1, M and C are numbers of misses and correct negatives for a given forecast of an event not occurring. Basic indices derived from the contingency table are summarized in Table 5.2.2.

Table 5.2.1. Contingency table for verification of dichotomous (yes or no) forecast. The number of hits, H indicates event occurrences that were correctly predicted. The number of false alarms, F denotes predictions of an event that was not observed. The number of misses M, indicate occurrences of an event that were incorrectly predicted. The number of correct negatives C, denotes correct forecasts of "no event". T denotes the number of total data.

		Forecast		
		Yes	No	Sub total
Observation	Yes	H (Hit)	M (Miss)	H+M
	No	F (False alarm)	C (Correct negatives)	F+C
	Sub total	H+F	M+C	T

Table 5.2.2. List of verification indices that could be derived from the contingency table.

Index	Definition	Formula	Characteristic	Range
ACC (forecast accuracy)	Ratio of correct forecasts to the total number of forecasts.	(correct forecasts) / (total forecasts) =(Z+H) / (N)	ACC is strongly influenced by the predominant number of correct "no event" forecasts, C	0.0 ~ 1.0, perfect score 1.0.
BIAS (bias score)	Relative frequency of predicted and observed phenomena, without regard to forecast accuracy.	(event forecasts) / (event observations) =(F+H) / (M+H)	If BIAS equals unity, then the predicted event frequency is the same as was observed, but it may or may not be located in the same time.	0.0 ~ ∞ (Infinitely great), perfect score 1.0.

Index	Definition	Formula	Characteristic	Range
POD (probability of detection)	Success of the forecast in correctly predicting the occurrence event.	(correct event forecasts) / (event observations) =(H)/ (M+H)	It is possible to score well on the POD by overforecasting the occurrence of yellow sand.	0.0 ~ 1.0, perfect score 1.0.
FAR (false alarm ratio)	Fraction of event predictions which were actually non-event	(false alarms) / (event forecasts) =(F) / (F+H)	It is used to check for the error that POD scores well by overforecasting the occurrence of event.	0.0 ~ 1.0, perfect score 0.0.
CSI (critical success index)	CSI takes into account both false alarms and missed events.	(correct event forecasts) / (event forecasts + observations) = (H) / (F+M+H)	CSI has the advantage of not being dominated by the no-events (C). CSI is biased toward data sets with higher event frequencies, making it a misleading statistic to use when comparing forecast skill score different regimes.	0.0 ~ 1.0, perfect score 1.0.

The CSI is evaluated at many national meteorological centers, but is not formally exchanged among nations. The CSI depends on the forecast lead time, rainfall classes, and target areas. The rainfall is a very local event, and the spatial distribution is difficult to get from spot observations. Standard procedures for its evaluation are not well established yet. Only a few national meteorological centers provide precipitation verification scores on the Internet. For instance, KMA website provides the ACC for its precipitation forecasts. The ACC for the official precipitation forecast with a threshold of 0.1 mm is around 85% for D+1, as shown in the upper panel of Fig. 5.2.4. This means that 15 forecast cases out of 100 cases miss the target. Another example from U.S. NWS is shown in the lower panel of Fig. 5.2.4. The threat score of 24-hr accumulated model precipitation exceeding 2 inches for D+2 have large fluctuations from 0.05 in summer to 0.3 in winter. The lowest score in the summer season from May to August is mostly caused by convective-type precipitation.

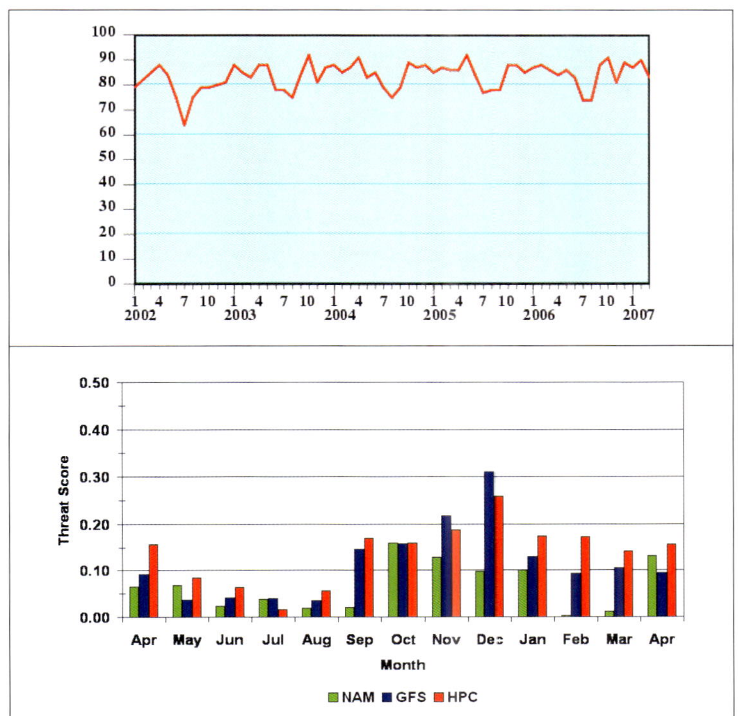

Fig. 5.2.4. Examples of verification for precipitation forecasts on the Internet. Accuracy (ACC) of D+1 official forecast from KMA for rain or no rain (upper), and critical success index (CSI or threat score) of D+2 model forecasts (NAM, GFS) and official forecasts (HPC) from the U.S. NOAA exceeding 2 inches of rainfall (lower). Similar information can be found for other weather elements, such as temperature, wind, etc. The images are available from http://www.kma.go.kr/weather/forecast/forecaetevalue_01.jsp and http://www.hpc.ncep.noaa.gov/html/hpcverif.shtml, respectively.

WMO website provides the standard verification scores of numerical models from national meteorological centers. Other websites also provide verification scores for the short and medium-range model forecasts including NCEP, ECMWF, and JTWC in Table 5.2.3. For the verification of dynamic seasonal forecasts, a few websites including APCC, ECMWF, NWS, and Development of a European Multimodel Ensemble system project (DEMETER) upload their verification scores.

Table 5.2.3. Some websites for verification of short and medium range forecasts. DYN, STAT, ENS, and OF refer to dynamical model, statistical model, ensemble prediction system, and official forecasts respectively. TC denotes verification for tropical cyclone track forecasts. SAT indicates verification with reference to satellite rainfall estimates.

Center	Type	URL (http://)
U.S. NWS	DYN OF	www.emc.ncep.noaa.gov/gmb/STATS/STATS.html www.emc.ncep.noaa.gov/mmb/research/meso.verf.html www.hpc.ncep.noaa.gov/html/hpcverif.shtml
	TC	www.nhc.noaa.gov/verification/
	STAT	www.nws.noaa.gov/mdl/synop/verification.htm slosh.nws.noaa.gov/verif/
	ENS	www.emc.ncep.noaa.gov/gmb/ens/verif.html www.emc.ncep.noaa.gov/mmb/SREF/VERIFICATION_32km/ new_html/system_48km_30day.html
DEMETER	ENS	www.ecmwf.int/research/demeter/d/charts/verification
ECMWF	DYN	www.ecmwf.int/products/forecasts/d/charts/medium/verification/ www.ecmwf.int/products/forecasts/d/charts/seasonal/verification/previous/
	TC	www.ecmwf.int/products/forecasts/d/charts/medium/tropcyclones/Verification/
KMA	OF	www.kma.go.kr/weather/forecast/forecaetevalue_01.jsp
UKMet	DYN	www.metoffice.gov.uk/research/weather/numerical-modelling/verification
BoM	SAT	www.bom.gov.au/bmrc/SatRainVal/sat_val_aus.html
WMO	DYN	www.wmo.int/pages/prog/www/DPFS/ProgressReports/index.html

Most of the measures discussed in this chapter are targeted for smoothed fields, and have certain limitations for noisy fields such as precipitation. Objective-based verification schemes are under development to incorporate pattern-matching technique in the verification of precipitation forecast (Gilleland et al., 2010).

5.3 Rule of thumb

It would be a pity to throw away meteorological information obtained from the

Internet, simply because its quality is not verified or if the information is not 100% perfect. Below are some tips that one could easily apply in evaluating the quality of weather forecasts. First, forecast accuracy decreases as its lead time increases. A forecast for today is more accurate than a forecast for tomorrow, which in turn is better than a forecast for the day after tomorrow. A forecast for D+3 is more accurate than a weekly advisory. The accuracy of computer-generated forecasts has a similar relationship with the lead time.

The anomaly correlation of state-of-the-art computer model forecasts for geopotential at 500 hPa surface ranges between 0.8 for D+5, and 0.6 for D+8 in Fig. 5.2.3. The predictability of precipitation decreases rapidly as lead time increases from D+1 to D+3. The Brier skill score and the threat score from the ECMWF and the NCEP models are presented in Fig. 5.3.1. The skill score of precipitation exceeding 2mm/day drops from 0.4 at D+1 to 0.3 at D+3 for the ECMWF model, and from 0.36 at D+1.5 to 0.18 at D+3.5 for the NCEP model. The threat score of precipitation exceeding 1 inch drops from 0.29 at D+1 to 0.23 at D+3 for the NCEP model, even though the threat scores have significantly improved since the early 80s.

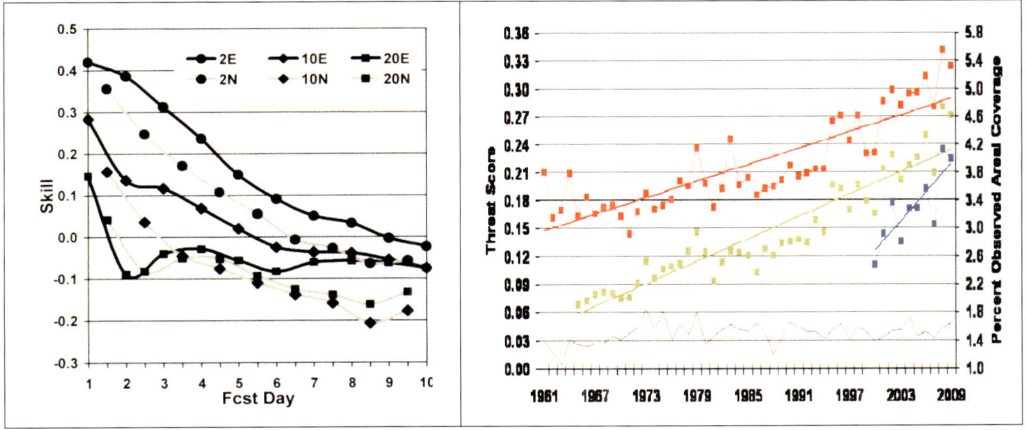

Fig. 5.3.1. (Left) Brier skill score for probability of precipitation over the United States from the ECMWF ensemble prediction system (black) and the NCEP ensemble prediction system (grey) between 1 Jan. 1997 and 31 Jan. 1998 with varying thresholds: 2 mm/day, 10 mm/day and 20

mm/day (Mullen and Buizza, 2001). (Right) Threat score for quantitative precipitation forecasts for 1 inch of precipitation at D+1 (red), D+2 (green), and D+3 (blue) hours, and percent areal coverage, i.e., the fraction of total forecast area to the area that actually received 1 inch or more precipitation from the U.S. NOAA Hydrometeorological Center (National Research Council, 2010). Details of ensemble prediction system and Brier skill score are explained in Chapter 6.

The relationship between forecast errors and lead time should always be considered in the application of weather forecasts. As the latest weather forecast update is the most credible, users should check the time of issuance. The ECMWF meteogram in Fig. 3.3.1 provides error bar or range of forecast values that indicate the degree of uncertainty of the corresponding forecast. Details of ensemble prediction system products will be discussed in Chapter 6.

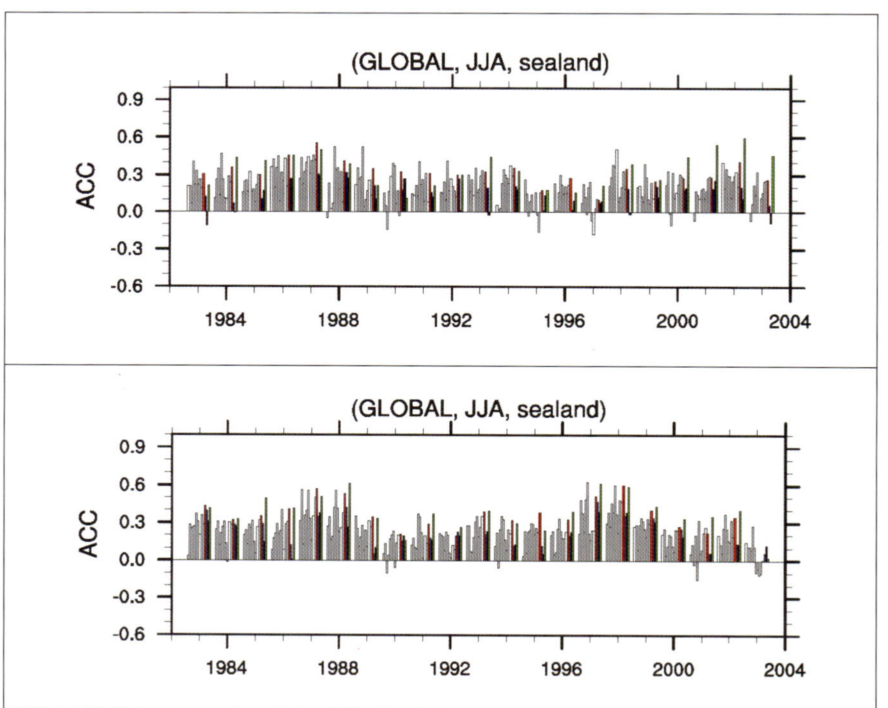

Fig. 5.3.2. Global accuracy, as measured with (5.2.4), of category forecasts (above/normal/below) for the following season (June, July, August) from a dozen of computer models shared at the APEC Climate Center (APCC) and their composites for temperature at 850 hPa (upper) and precipitation (lower).

The accuracy of seasonal forecast is much lower than that of medium-range forecasts. As seen in Fig. 5.3.2 for the dynamic seasonal forecasts from the APEC Climate Center (APCC), the accuracy (ACC) for temperature at 850 hPa and precipitation is only about 0.3 with large variability across models and seasons. Furthermore, the forecast accuracy highly varies from region to region. The explained variability of the hindcast relative to the observed variability is particularly useful for evaluating the skill or uncertainty of seasonal forecast at any locality. The Mean Squared Skill Score (MSSS) is defined as one minus the ratio of the mean squared error of the forecasts to the mean squared error of climatology.

As the monthly or seasonal forecast is less accurate than the weekly forecast, their departure from the climatology is predicted in terms of probability. For instance, the APCC seasonal forecast for June, July and August in Fig. 5.3.3 shows moderately high confidence over tropical continents and some confidence in Europe, Central Asia, and the western United States. However, it never exceeds 0.4, which means that about 60% of climate variability cannot be explained by the forecasts. Those uncertainties should be considered when applying seasonal forecasts in weather- and climate-sensitive decision-making areas.

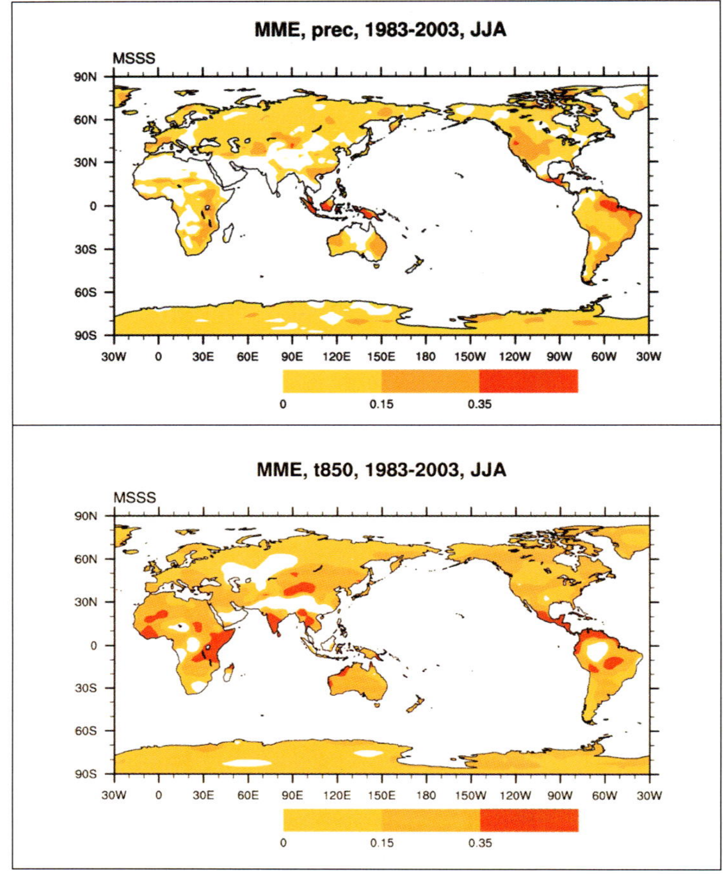

Fig. 5.3.3. Mean squared skill scores of hindcast during 1983~2003 for the precipitation (upper) and temperature at 850 hPa (lower) during June, July, and August, based on the ensemble mean of a dozen of computer models shared at APCC.

Second, predictability varies with the spatial scale of atmospheric motion. The atmosphere can be viewed as a combination of multiple scales of motions, like the toothed wheels of a clock. Troughs or ridges in an upper-level chart represents large-scale motion with wavelengths of thousands of kilometers. These systems are predictable up to 5 days, as evidenced in the upper panel of Fig. 5.3.4. Very long waves with wavelengths of 10^5 kilometers or more can be predicted for 10 days. Particularly, predictability extends out to several weeks when a dynamically stable system such as blocking persists. By contrast, short wave troughs with a wavelength of hundreds of kilometers have very limited predictability - a few days. Frontal rain organized by low depression is more predictable than sporadic thunder showers. The predictability of individual storm cells is less than 30 minutes, which have a characteristic length scale of 5km or less. Generally speaking, the smaller the size of a system the poorer the predictability, and accordingly, the shorter the forecast lead time. By the same reason, the model fields in the upper troposphere are more predictable than those in the lower troposphere where complex terrain and surface texture force small-scale motions. The anomaly correlation of a D+5 model field near the surface is less than 80 % of that at 10 hPa, as demonstrated in the lower panel of Fig. 5.3.4. The anomaly correlation of the surface field drops rapidly with time, and amounts to 30 % of that of upper level at D+10.

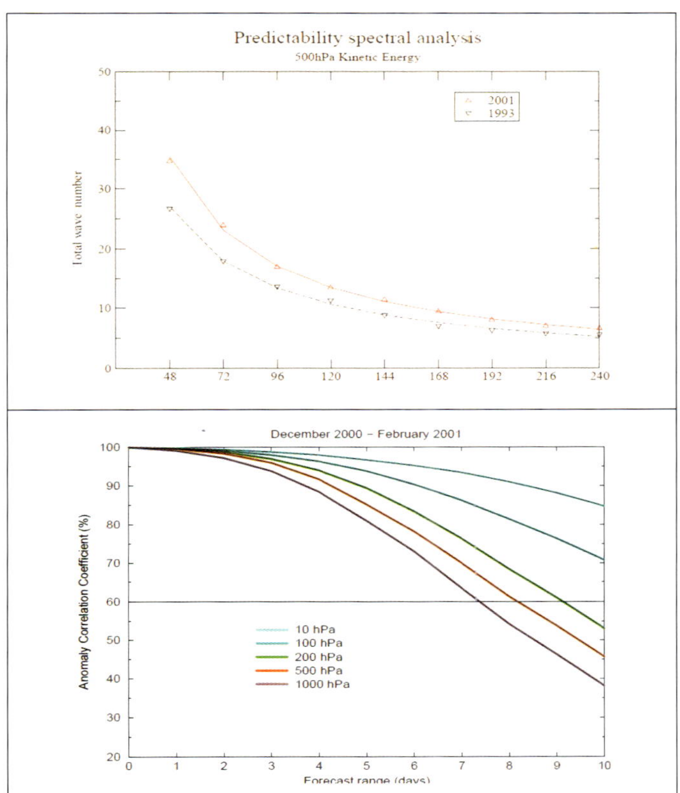

Fig. 5.3.4. Predictability of ECMWF model fields with varying horizontal scales in terms of forecast

periods over 0.6 (upper), and vertical levels in terms of anomaly correlation coefficient (lower) over the Northern Hemisphere winter (Persson, 2001).

Third, predictability differs from variable to variable. Precipitation is the most unpredictable among the weather elements (Fritsch and Carbone, 2004). Rain or snow results from multiple interactions among various scales ranging from raindrops of a few millimeters to mid-level cyclones spanning a few thousand kilometers. Similarly, the variables directly related to the moisture field are also difficult to predict. Table 5.3.1 presents the current skill of the ECMWF model for the various weather attributes. Cyclone life cycle and fronts can be predicted up to D+5. The table confirms that precipitation and clouds are less predictable than temperature and wind. Model precipitation shows reasonable skill up to D+3 at the most.

Table 5.3.1. Current skill of the ECMWF model (Persson, 2001).

Feature	<D+3	D+3 to D+5	D+5 to D+7	D+7 to D+10
Hemispheric flow transitions	Excellent	Excellent	Good	Some skill
Blocking creation and breakdown	Perfect	Good	Fair	Low skill
Cyclone's life cycle	Perfect	Fair	Low skill	-
Fronts and 2nd development	Good	Fair	-	-
Temperature and wind	Very good	Skill in daily extremes	Skill in 5~10 day mean	
Precipitation and mean clouds	Good	Some skill	Some skill in precipitation 5~10 day accumulated values	

Even though temperature and wind are more predictable than precipitation, their accuracy falls over coastal zones and mountain areas where land-sea breeze or mountain-valley breeze cause large fluctuations in local weather. Thus it is wise to put more effort into the interpretation of large-scale dynamic variables first, then proceed with caution to model precipitation and the local wind.

Lastly, intensity forecasts or quantitative forecasts are less predictable. Quantitative precipitation forecast is a good example. Based on statistics of precipitation forecasts from ECMWF and NCEP in Fig. 5.3.1, Brier skill score for probability of precipitation with a threshold of 20 mm/day is almost null, while that of 2 mm/day ranges from 0.1 to 0.4 depending on lead time. The difficulty in forecasting of tropical cyclone intensity will be elaborated in Chapter 8.

Consistency check

Users who have some knowledge on meteorology can make use of the relationship among the weather elements. In many cases, a single website provides many forecast charts on the same weather event. For instance, various prognostic charts are displayed on the NCEP website, including temperature, humidity, geopotential, and wind at each standard pressure level. One could check if the relationship among them is consistent with the basic rules of atmospheric physics such as thermal wind relationship and baroclinic wave development. Some websites upload auxiliary forecast charts along with the basic charts. For instance, the vertical velocity field, vorticity field, and moisture flux are found on the websites of KMA, NCEP, and NOGAPS. A sample layout of model variables from NOGAPS (US Navy) is presented in Fig. 5.3.5. Regions with upward motion are favorable for clouds and precipitation. In a synoptic sense, the upward motion field is found downstream of an approaching upper-level trough or over a region where warm air advection occurs, as discussed in Chapter 4, reflecting the relationship among precipitation, vorticity at 500 hPa, and temperature at 850 hPa. If a forecast chart poorly represents the physical relationship, its reliability is questionable. It is recommended to consult with a professional meteorologist for the assessment of the physical relationship among the model fields.

	PRODUCT	all	all	all	all	all	all	all	all	all	all	all	all	all	all	all	all	
TAU		000	012	024	036	048	060	072	084	096	108	120	132	144	156	168	180	Loop
all	300hPa Heights [m] and Isotachs [kts]	●	●	●	●	●	●	●	●	●	●	●	●	●	●	●	●	●
all	500 hPa Heights [m] and Rel. Vort [10-5 s-1]	●	●	●	●	●	●	●	●	●	●	●	●	●	●	●	●	●
all	700 hPa heights ; Rel. Hum. [%] ; Vertical Velocity [Pa/s]	●	●	●	●	●	●	●	●	●	●	●	●	●	●	●	●	●
all	850 hPa Temperature [C], winds [kts] and Rel. Hum. [%]	●	●	●	●	●	●	●	●	●	●	●	●	●	●	●	●	●
all	1000-500 Thickness [dm] and Sea Level Pressure [mb]	●	●	●	●	●	●	●	●	●	●	●	●	●	●	●	●	●
all	Previous 12-hr Precipitation Rate [mm/12hr] and Sea Level Pressure [hPa]	●	●	●	●	●	●	●	●	●	●	●	●	●	●	●	●	●
all	FNMOC Wave Watch 3 Sig Wave Heights [ft] ; Over Ocean Sfc Winds [kt]	●	●	●	●	●	●	●	●	●	●	●	●	●	●	●	●	●
all	30K-38K ft CAT	●	●	●	●	●	●	●	●	●	●	●	●	●	●	●	●	●
all	22K-30K ft CAT	●	●	●	●	●	●	●	●	●	●	●	●	●	●	●	●	●
all	Magnetic Ditch Headings (DAF)	●	●	●	●	●	●	●	●	●	●	●	●	●	●	●	●	●
all	Altimeter Setting	●	●	●	●	●	●	●	●	●	●	●	●	●	●	●	●	●

Fig. 5.3.5. A layout of model variables appeared on the NOGAPS (U.S. Navy) website. Users can select one of 13 theaters around the world. The model forecasts up to 180 hours appear on the screen all together when one chooses a forecast variable and a forecast time from the chessboard menu. The image is available from http://www.usno.navy.mil/FNMOC/meteorology-products-1 or https://www.fnmoc.navy.mil/wxmap_cgi/index.html.

The sequence of forecast runs helps one assess the internal consistency of forecast charts. If this morning forecast differs markedly from the last night forecast, it is less than reliable. Inconsistency indicates a weakness in the data assimilation system that provides the initial condition for a computer model. The same holds true for the sequence of weather forecasts. Figure 5.3.6 presents a sequence of model runs at 12-hr intervals. The consistency of precipitation pattern is moderately poor after D+4 when the forecaster's confidence is low. While the latest run indicates slow migration of a stationary front and intensification of rainfall band in the target region around D+6, previous runs predicted earlier decay of the frontal system followed by a weak secondary system approaching the domain from the west. The problem of consistency often occurs when pressure patterns abruptly change from one weather regime to another, or when a flow regime is controlled by weak thermal forcing over a data-sparse area as in the case of a tropical depression over the ocean.

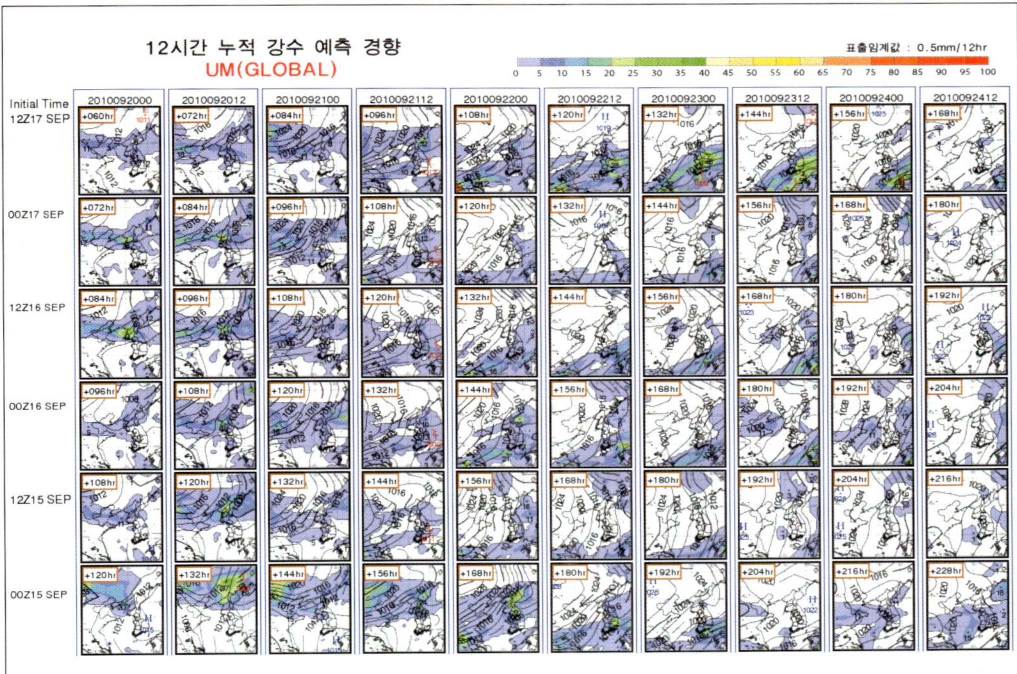

Fig. 5.3.6. Consistency among model forecasts from a sequence of the UM model runs with a grid spacing of 40km and 50 levels valid 20~24 September 2010 with 6 initial conditions during 15 ~17 September 2010. Each column represents different model forecasts for mean sea level pressure (solid line) and precipitation amount (color) with the same valid time as indicated in the top row. Source: KMA.

5.4 Comparison with current weather

The quality of model forecasts discussed in the previous sections is based on averaged properties over a large domain during long periods to assure statistical significance. It does not tell us the specific characteristic of a given model at any instant or location of interest. Thus careful monitoring of model behavior in light of onsite observations is always helpful in capitalizing on the model output in a given forecast theater.

Forecasters regularly assess model performance based on daily episodes; this review is an integral part of the forecast process. In fact, forecasters at operational centers are busy comparing computer model forecasts with the ongoing weather to choose the best, and to make necessary adjustments to ensure model forecasts are consistent with the latest changes in the observations. A model analysis at initial time is often regarded as a proxy to observation, and is used to verify the model forecast. However, the model analysis is not free from flaws in the model, as the model forecasts are used as a background guess to derive the model analysis. This is particularly true for model analyses over data-sparse areas.

Satellite imagery and radar loop provide rich information on the ongoing cloud systems and precipitation bands. They are instrumental in demonstrating the discrepancy between the model forecast and the current state of the atmosphere. Extra data processing, however, is required to convert satellite radiances or radar reflectivity into model variables. Models face spin up problems for clouds and precipitation in the nowcasting range, i.e., 0~6 hours, which arise mostly from the incompleteness of moist physics and uncertainty of the initial field for moisture and various types of hydrometeors in the model. The benefit of assimilation of radar or other nowcasting information lasts only during the first few hours of a model run. Thus it is misleading to pay too much attention on any change in very short-term model precipitation between successive model runs. To access model bias and characteristics, it is more useful to compare verifying radar reflectivity with model precipitation from 12 hours and onwards.

Using conceptual models, certain dynamic features of model forecasts can be indirectly compared with the observations. The conceptual model of fronts in Chapter 4 is useful for matching the model precipitation with the observed temperature gradients and wind shear. The conceptual model of Tropical Upper Tropospheric Trough (TUTT) is a good indicator for the analysis of upper-level divergence that supports tropical storm development with large-scale ascent motion. The positive isentropic potential vorticity (IPV) anomaly on the dynamic tropopause, defined by the geopotential height contours of 1.5 IPV unit grossly matches with the active forcing region in the upper level according to the framework of so-called PV-thinking concepts (Hoskins et al., 1985). It is manifested by the dark edge of a dry slot in the satellite image of the water vapor channel, which will also be discussed in Chapter 11. The identified phase error of model IPV anomaly as such can be used to correct the subsequent forecast phases of surface cyclones or fronts in the model.

6. Probability and Application

6.1 Deterministic chaos

Clocks and clouds represent deterministic and stochastic systems, or closed and open systems respectively. The needles and toothed wheels of a clock move under definite rules. They do not interact with the environment as long as the battery is charged. Their movement is transparent to everybody, and so is the time. We can make an appointment with others because time is predictable.

On the other hand, clouds are evolving and changing all the time. A single cloud has an almost infinite number of water drops or ice particles. Rules governing the interaction among those particles are not completely known. Clouds exchange heat, moisture and momentum with the environment. The current state of clouds is uncertain, and so is their future state. Most weather systems such as low depressions, tropical cyclones, heavy snow bands and flash flood systems are open systems. They are in some sense as uncertain as clouds.

The limitation of predictability in weather forecasting is rooted in the chaotic nature of the atmospheric system. Dr. Edward Lorenz, a meteorologist, discovered chaotic phenomena in a simplified dynamical system that consists of three time-dependent nonlinear equations. A state in the system is defined by three parameters. Let us consider states A and B in the three-dimensional phase space; the two states are close to each other within the range of observation errors. Two very similar states at an initial time may evolve into different states A' and B' respectively. The distance between the two states is considered a measure of error or uncertainty. If the distance grows within a finite time to the random difference of any pair of observed states, the predictability of the system is lost, as it is difficult to determine which one is the true state. This happens when the system is very sensitive to initial conditions due to the inherent dynamic instability, as demonstrated in the right panel of Fig. 6.1.1.

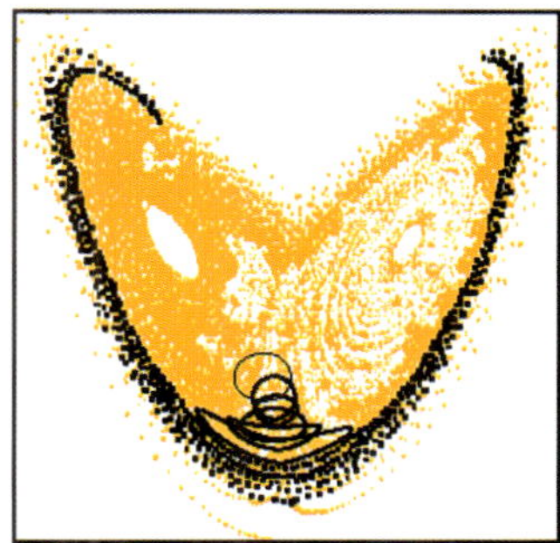

Fig. 6.1.1. Phase-space evolution of an ensemble of initial points for three different sets of initial conditions on the Lorenz attractor in two-dimensional parameter spaces, as it appears in numerical experiments on a nonlinear system with three equations (Palmer, 2003). Predictability is a function of the initial state. Each point represents a state in the system. The sequence of points corresponds to the evolution of the state. The difference between the pair of states, slightly displaced at an initial time, would eventually be so magnified and distorted that each of them would be attracted to opposite sides of the butterfly wings. This dependency is illustrated by showing the growth of an initial isopleth of an idealized probability distribution function at three different positions on the Lorenz attractor. In the first position, there is little spread and predictability is large. In the second position there is some spread and predictability is moderate. In the third position, initial growth is large and the resulting predictability is small. If both wings represent climate regimes A and B, a small difference in initial states may lead to completely different climates. The predictability of weather is severely restricted by the sensitivity of the nonlinear system to the initial condition.

The atmosphere is understood as more chaotic than the Lorenz system in at least two aspects. First, the atmospheric state is so complex that its evolution can hardly be expressed by a finite number of parameters. Second, the atmosphere is an open system that exchanges energy with the land surface and biosphere, ocean floor, and polar ice sheets. In addition, there are many unknowns in the physics of the atmosphere, particularly regarding interactions between large-scale motion and small-scale motion, and chemical reactions among various constituents including greenhouse gases and aerosols. The current observation network is grossly insufficient to resolve the small-scale motion field. Accordingly, an analysis of the initial state based on the available observation network is a poor approximation of

Mother Nature. Error in the initial condition is unavoidable and the errors embedded in small-scale motion tend to grow faster than those in large-scale motion. The predictability of the atmospheric state is severely bounded in a finite time due to chaotic behavior.

As discussed in section 3.3, only the likelihood of thunderstorm can be predicted a few hours ahead. The uncertainties of time and energy compete with each other. It seems difficult to forecast accurately both the intensity and the timing of heavy rainfall. One sacrifices the accuracy of timing for maximum development, if one wants to improve the estimate of a storm's growth rate.

Large-scale motion of the atmosphere can be predicted a week in advance with the current science and technology. The maximum limit of predictability of a weather forecast is about two weeks according to chaos theory. However, the predictability of a time mean state or climatology may extend beyond two weeks in a favorable weather regime, where atmospheric motion responds to slowly varying boundary conditions such as ocean temperature and land surface features.

6.2 Probability forecast

The limitation of predictability means that uncertainty is inherent in weather and climate forecasts. The concept of uncertainty is reflected in forecast terminology such as "possibility of rain", "occasional showers", "chance of rain", "sporadic," and "patch fog." Users will interpret these terms differently depending on their personal preferences and interests. The best way to represent uncertainty is to quantify it in terms of probability between 0 to 100%. For instance a 70% chance of rain in Washington tomorrow means that 70 cases out of 100 cases would record rain under the same meteorological conditions for tomorrow. Let us consider another example from long-range forecasts. A forecast saying "40% warmer than normal, 30% normal, and 30% below normal" means there will be a slightly higher chance of warmer conditions than normal next month.

Despite the risk of forecast failure, the general public tends to prefer definitive information over ambiguous forecasts. Most weather forecasts for the public are written in definitive terms, such as "rain", "fair", so on. A categorical forecast with clearly defined wording is nothing but a forecast specifically tailored to the targeted user group. Suppose that a forecaster, projecting a 20% chance of rain for tomorrow, expresses the forecast simply by saying "rain tomorrow." The forecast could be good or bad, depending on the client's sensitivity to weather information. Let's say a person is involved in a business sensitive to weather conditions like rain. Even a small chance of rain may cause critical damage to the business. In such cases, the categorical forecast would be appropriate for the client. On the other hand, there are other businesses with little sensitivity to rain. In such cases the same categorical forecast may cause unnecessary financial cost from over preparations. This client should have discarded the categorical forecast, if a probabilistic forecast were given instead. A categorical forecast turns out to be a biased forecast for a client whose interest does not exactly match the forecaster's. The only way to satisfy every user group is to provide them with probabilistic forecasts to be interpreted at their discretion, considering their own sensitivity to weather conditions. The user could then recruit a specialist to interpret the probabilistic forecast into actionable information for the business concerned.

Probabilistic forecasts are frequently found on the Internet. The major national meteorological centers provide probabilistic forecasts for weather elements such as the occurrence of rain or thunderstorms, and for continuous variables such as minimum and maximum temperature and precipitation exceeding certain threshold values. Digital forecast websites in Fig. 3.3.2 also provide various statistical forecasts including probability of quantitative precipitation and conditional probability of severe weather events:

http://www.nws.noaa.gov/mdl/forecast/graphics/MET/gifs/MET.PQPF324.000.gif

http://www.weather.gov/mdl/forecast/graphics/MET/gifs/METTSVR.USVR24.000.gif

There are two approaches to derive probabilistic forecasts: one may use either a deterministic model or an ensemble prediction system (EPS). In the first approach, the statistical relationship between the time series of model forecasts and observations is used to identify systematic biases in model forecasts. Random errors from various defects in model physics and data assimilation filter out using statistical average. A prime example of this approach is model output statistics (MOS). MOS enables the prediction of surface weather elements in continuous or binary form and the probabilistic distribution of their occurrences. MOS also exploits the predictability of large-scale parameters in the free atmosphere to correct model bias near the surface. MOS fails when a model poorly simulates the evolution of an ongoing synoptic system. Precipitation is the least predictable using MOS, and subjective adjustment by the forecaster is indispensable.

In normal situations, humans add little to MOS forecasts for surface weather elements including temperature. This point was confirmed by a study conducted at the Meteorological Service of Canada. However, bias associated with fast-evolving systems such as convection and explosive cyclogenesis may not be properly reflected in MOS. Training samples for MOS mostly cover short periods not exceeding three years. MOS cannot explain the outliers from extreme weather conditions. Users must be aware of the limitation of those statistical forecasts on extreme weather cases where forecast guidance is desperately needed. Another drawback of MOS-based probability forecast is the neglect of in-situ flow dependency. The samples used for the derivation of bias characteristics contain various types of synoptic flows in an aggregate form that are not necessarily similar to the given synoptic flow. The second approach based on EPS is designed to avoid this limitation in MOS and will be discussed in the next section.

6.3 Ensemble and forecast scenarios

An ensemble of model forecasts from EPS can be used for probabilistic forecasts. The approach samples over multiple simulations from slightly different initial states or physical parameterization schemes under a given large-scale flow situation. By doing so, it considers the in-situ flow dependency in assessing uncertainty if a sufficient number of ensemble members were used for the analysis. This is the very property missing from MOS and conventional approaches.

An ensemble literally means a group of things or people considered as a whole. A chamber ensemble is a group of musicians playing violin, viola, and cello. When listening to the opera Aida we are impressed by the integrity and harmony from the ensemble of voices. Like the chorus filtering out the individual noise, the ensemble of forecast charts percolates forecast errors. The heart of EPS is to distill biases in forecast information by averaging with weight over ensemble members. An ideal ensemble consists of forecast members with independent error characteristics in the sample space.

One of the advantages of the ensemble forecast is to quantitatively measure the degree of uncertainty associated with the forecast. When the forecast fields in an ensemble resemble each other, the uncertainty of the forecast may be reduced. That is, the forecaster can make a decision with more confidence. If the forecast fields in the ensemble diverge, the forecaster loses confidence and the uncertainty of the forecast increases. Roughly speaking, large spread means low predictability. However, a small spread does not necessarily guarantee the reliability of a forecast. In the worst case, all the solutions drift in the wrong direction. When a great storm in 1987 swept over the southern part of England, most numerical models failed to predict its path correctly. A similar problem occurred in the operational models for the prediction of the Presidents' Day storm in February 1979 and the Inaugural Day windstorm in January 1993 in the United States.

The classic approach with EPS is to generate slightly different initial conditions by adding perturbations to the analysis that are comparable to observation error. The perturbed initial conditions then feed into the model, and yield different forecasts. The sensitivity of initial conditions can be measured by the degree of spread among the forecasts.

The most familiar ensemble is the collection of forecast charts in use at major operational centers, assuming that computer models and their physical processes are all independent of each other. This approach is called as a "poor person's ensemble," as it requires little cost for its implementation. It has been widely used at operational centers for short and medium range forecasts. One may construct such an ensemble by collecting 500 hPa forecast charts concurrently on the KMA, JMA, ECMWF websites. Another example can be taken from the track forecasts on the KMA, JMA and JTWC websites. As the sample size is limited in a poor person's ensemble, such ensembles can be used only for the qualitative measure of uncertainty, while the forecasters could spend more time to develop down-scaled scenarios for extreme weather events.

To take errors of both initial conditions and models into account, the THORPEX Interactive Grand Global Ensemble (TIGGE) International project suggests using an ensemble of ensembles or grand ensemble. Figure 6.3.1 illustrates the most complex form of an ensemble that combines both classical EPS and poor person's ensembles.

The quality of EPS guidance is sensitive to the ensemble construction. Each forecasting center uses a different model and different initial analyses for the production of forecast charts. The first step before using EPS products is to check how the ensemble is composed. If the EPS does not take account physics uncertainty, it underestimates forecast uncertainty. The ECMWF EPS introduces perturbation of moist physics over warm oceans. It is better at representing forecast uncertainty in tropical cyclone intensity than other EPSs that only consider the perturbations in the initial states.

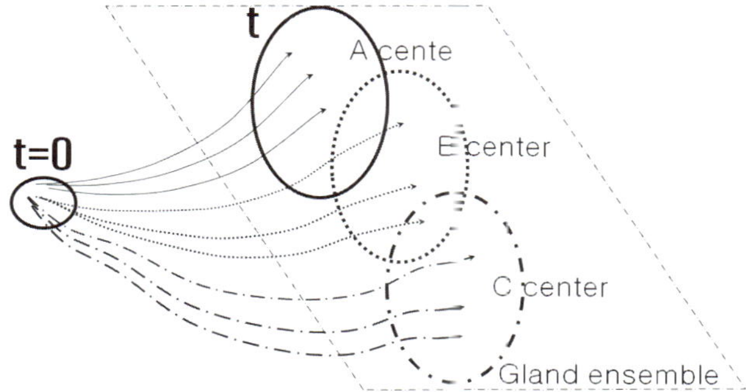

Fig. 6.3.1. Schematic diagram for a grand ensemble in the phase space. Each arrow indicates the trajectory of a state projected into the future by the computer model during the period (0, t). Centers A, B, and C provide their own ensembles, which are combined to form a grand ensemble.

In principle, for every increase N in the number of EPS members, the information contents of EPS increase by N times over the corresponding deterministic forecast. The quality of each member should be monitored using the metrics discussed in Chapter 5 to construct better ensembles, while the physical reasons for the discrepancy among members are explored. To digest all the available EPS output is a formidable task for the forecaster who has to produce forecasts and warnings in a very limited time frame. The heart of EPS application is how to grasp major features from the huge amount of data. The simplest way is to choose the most proper forecast from the ensemble that fits best the actual weather at the beginning of the forecast period. A more sophisticated way is to use the weighted average over the ensemble of forecast charts with reference to the latest performance of each forecast. The typical weight is inversely proportional to forecast error. The mean over the ensemble, i.e., ensemble mean, is often used as a substitute for deterministic forecasts, and has proved to be more accurate than the latter.

Many websites provide a variety of graphics for ensemble forecasts. Some websites are listed in Table 6.3.1. The box and whisker diagram in Fig. 3.3.1 helps us understand the spread of forecast values in the time series. The top whiskers extend to the highest value in the ensemble, while the bottom whiskers extend to the lowest value. The top and bottom of the box give the lower and upper limits of the middle two quartiles. The median is also shown in the box. As the box stretches, the uncertainty increases. Any point where a median value is displaced from the middle indicates that the ensemble members are not distributed evenly. If the median is above the mean, the lower half of the ensemble is more spread out than the upper half.

A spaghetti plot for geopotentials at 500 hPa in Fig. 4.2.4 is another example of an ensemble forecast. It expresses a bundle of contours, each of which corresponds to a certain value of geopotential height from the ensemble. The distance between ensemble members is measured by the spread of the contours. The uncertainty of the forecast increases as the spread increases and vice versa.

Table 6.3.1. Some websites for probabilistic guidance on atmospheric circulation based on ensemble prediction systems. For more information, contact GDPFS Expert Team on Ensemble Prediction Systems (ET-EPS) under CBS - OPAG/ WMO.

Meteorological Center	URL (http://)
CMC (Canada)	www.weatheroffice.gc.ca/ensemble/index_e.html
ECMWF	www.ecmwf.int/products/forecasts/d/charts/medium/eps/
NCEP (USA)	www.emc.ncep.noaa.gov/gmb/ens/ www.emc.ncep.noaa.gov/mmb/SREF/SREF.html

The probability of exceedance plots is also commonly used to estimate the likelihood of extreme weather associated with precipitation, temperature and gustiness exceeding a certain threshold. The probability is calculated from the ranked histogram by taking account of the number of ensemble members that exceed the chosen threshold and dividing it by the total number of members in the ensemble.

Many other diagrams help interpret ensemble forecasts. A comprehensive guide for the interpretation of ensemble forecasts is found in the COMET module: http://www.meted.ucar.edu/nwp/pcu1/ensemble/.

EPS based on initial perturbations often uses the low-resolution version of a deterministic model, while high-resolution model products are available at the same time. The EPS from ECMWF is a good example, as illustrated in Fig. 3.3.1. Under such circumstances, it would be helpful to digest first the prognosis from the high-resolution model, then return to the EPS with the low-resolution model to check the possibility of extreme events occurrence.

Even though the contemporary computer technology cannot support high-resolution EPS operation, model climatology provides additional insight on the degree of extremes or abnormality. Figure 6.3.2 illustrates the extreme forecast index (EFI) from ECMWF, which estimates the departures of EPS distribution from the model climatology. The EFI takes values between -1 for all EPS members forecast values below the climatological minimum, and 1 for all EPS members forecast values above the climatological maximum. The index helps alert forecasters in advance to the possibility of extreme weather occurrence. In Fig. 6.3.2, the model consistently forecast abnormally cold temperature over the Korean Peninsula from D-5 with an EFI value below -0.8 which nicely matched with the observed minimum temperature of 6~8 degrees below climatological normals.

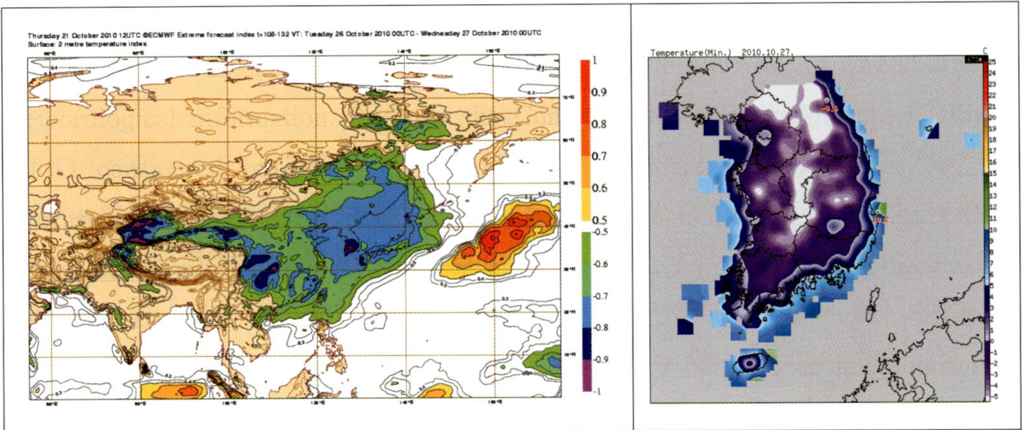

Fig. 6.3.2. Extreme forecast index (EFI) produced by ECMWF. D+5 forecast of 2m temperature (left) targeted for 27 October 2010, with verifying minimum temperature (right). The temperature minimum on that day is below freezing level inland, which amounts to -7 ~ -9 °C lower than the climatological normal in the central part of the Korean Peninsula (source: KMA). The EFI image is available from http://www.ecmwf.int/products/forecasts/d/charts/medium/eps/efi/efi_2t/.

The accuracy of probabilistic forecasts can be measured using the Brier skill score (BSS). The score is derived from an averaged sum of squares of differences between the forecasted probability and the observed.

$$BSS = \left[1 - \frac{\sum_{i=1}^{N} \left(F_i - O_i \right)^2}{\sum_{i=1}^{N} \left(C_i - O_i \right)^2} \right] \times 100 \qquad (6.3.1)$$

where N is the number of points in the domain. F_i is the forecast probability at i's point. O_i is the corresponding observed value at the same point. O_i has a value of 1 if the event occurs, and 0 otherwise. C_i is the climatological probability, or frequency of the event occurring at the i's point. The BSS score has values ranging from 0 to 1. The perfect forecast has a zero value, and the null forecast has a value of 1.

6.4 Decision-making under uncertainty

The value of probability forecasts, whether they be derived from MOS or EPS, is realized only when these are translated into categorical forecasts that lead to affirmative or negative actions in responding to weather-sensitive activities. Business productivity in most cases becomes sensitive to weather only when the value of weather attributes reach a certain threshold. Let us consider the case of a manager who works at a construction site. Activities such as concrete pouring is very sensitive to rain. The manager has to decide whether to proceed or stop working, faced with an expected 25% chance of precipitation for tomorrow. The first thing the manager should do is to evaluate the weather-sensitivity of his business: the ratio between the loss (L) from rain and cost (C) for protection against rain. The cost is large when the field work is hampered by rain. Furthermore, the manager has to pay the daily wages for the employees. Let's say that the total loss is 1 M dollars, and that the expenses for the opportunity loss are about 0.1 M for canceling the job. The ratio of cost to loss is 1:10. The expected value for the expense is 0.25×0.1 M+ 0.75×0.1 M = 0.1 M for taking measures against rain, and 0.25×1 M+0.75×0 M = 0.25 M for taking no measures against rain. The expense of taking measures is less than that of taking no measures. If a similar situation should occur many times, it would be wise for the manager to take measures against rain as long as the forecast of rain probability exceeds 25%.

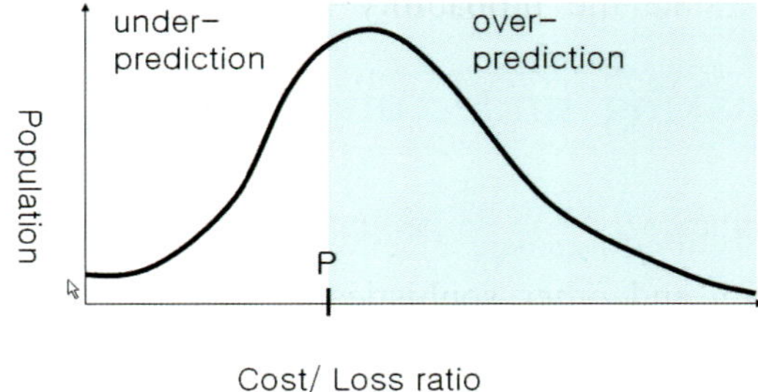

Fig. 6.4.1. Bias of a categorical forecast given the probability forecast P for a client represented by the cost/loss ratio. The client with a cost/loss ratio exceeding the forecast probability P is over-biased. Similarly, the client with a cost/loss ratio below P is under-biased.

Figure 6.4.1 illustrates bias of a categorical forecast of an event occurring given the probability forecast P for client represented by the cost/loss ratio. One would be wise to protect against rain if the probability of precipitation is greater than the cost/loss ratio. Otherwise it would be beneficial not to take any action at all. The application of probability forecast can be easily extended to other variables such as maximum and minimum temperature.

In our context, each business can be uniquely defined by certain threshold values. For instance, KMA issues heavy rain warning if more than 70 mm of rainfall is expected in the next 6 hours. The criteria is reasonable for taking action to protect property and lives against inundation of farming lands and facilities, landslides and flooding of low-lying areas. However, the same criteria are not well-suited for decision-making in the building industry, chemical plants, or airline control towers.

Certain types of hot breads sell better when the minimum temperature drops by more than 10 degrees in a day. Certain vegetables freeze if the air temperature drops below -2℃ and persists for more than 12 hours. Certain outdoor sports can tolerate precipitation if the intensity is less than 5 mm per hour. A city's demand for electricity abruptly rises for cooling if the maximum temperature goes beyond 35℃ in summer. It is the users that determine the threshold value critical to their business, and that translate the probability forecasts into action. The question is how reliable a given probability forecast is. Even rationally driven decisions may go wrong if a weather forecast is not reliable enough in the given situation. Forecast quality should be constantly monitored in terms of the metrics discussed in sections 5.2 and 6.3 and other sophisticated metrics.

It is noted that the probability forecast is only a part of the information for decision-making. Forecasters generally are not aware of business-specific information. In addition, the threshold value and cost/loss ratio for the relevant weather attribute reflects the user's interest, and is not necessarily available to forecasters. On the other hand, the accuracy of a probability forecast is not well understood by users. Thus, networking among forecasters and users is essential to share the needed information, and properly apply probability forecasts for decision- making.

Part Ⅲ. Large-scale Patterns

7. General Circulation and Long-range Forecasts

8. Cyclone Forecasts

7. General Circulation and Long-range Forecasts

The atmosphere can be viewed as an orchestra with multiple scales of motions with different lifetimes. The flow pattern in the example of Fig. 4.2.3 is organized by synoptic waves whose lifetime ranges from a few days to a week with horizontal scales of few thousand kilometers. The planetary-scale waves with much longer temporal scales modulate the synoptic waves.

State-of-the-art computer models capture the synoptic scales of motions, but face certain limitations in representing mesoscale details of a few dozens of kilometers or less. Thus the primary focus should be put on the synoptic scale or longer, when one interprets computer-generated charts from major meteorological centers. Understanding general circulation helps users develop a conceptual framework for deriving regional forecasts from computer model outputs.

7.1 Latitudinal dependency

Differences in temperature, density, and air pressure from the poles to the equator are the most distinct features of the global climate. These differences mainly stem from the seasonal variation of the solar zenith angle that the beam of sunlight makes with the plane surface of the earth. The earth receives more heat in the summer hemisphere as the sunlight is directed toward the earth in a more upright direction. Similarly, the earth takes less heat in the winter hemisphere.

The geographical distribution of the ultraviolet index in Fig. 7.1.1 reflects the seasonal march of solar radiance. The UV index values exceed 10 near the tropics all the time, and 7 during summer in high-latitudes. The daily index, however, may vary with cloud cover and stratospheric ozone density. The value may go up to 15 units in high mountain areas.

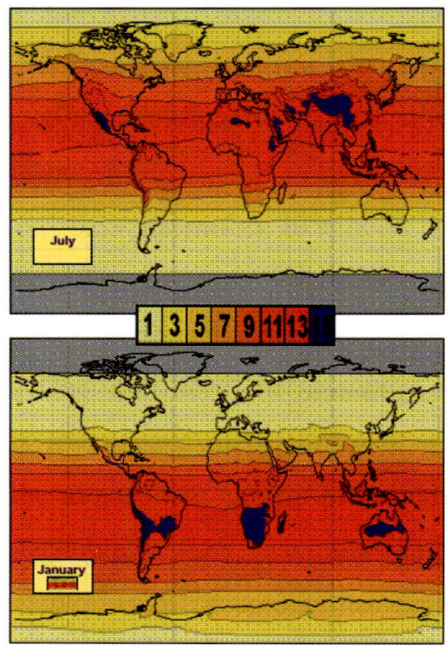

Fig. 7.1.1. Global distribution of the UV Index for clear sky at noon in July (upper) and January (lower) (Vanicek et al., 1999).

There are transitory seasons between summer and winter in mid-latitudes. It is hot and humid all the time in the tropics. In between are the moderately hot and dry regions, the subtropics, where most of deserts are located. The Arctic and Antarctic regions are very cold.

The latitudinal differences in temperature cause air to move from warm to cold regions, as illustrated in Fig. 7.1.2. These currents help transport heat from the equator to poles. Thermally direct circulation, i.e., warm air rising and relatively cool air sinking, occurs in the Hadley cell that extends from the equator to the subtropical region around 30 degrees in latitude. Reverse circulation occurs between 30 to 60 degrees in latitude where synoptic waves are dominant.

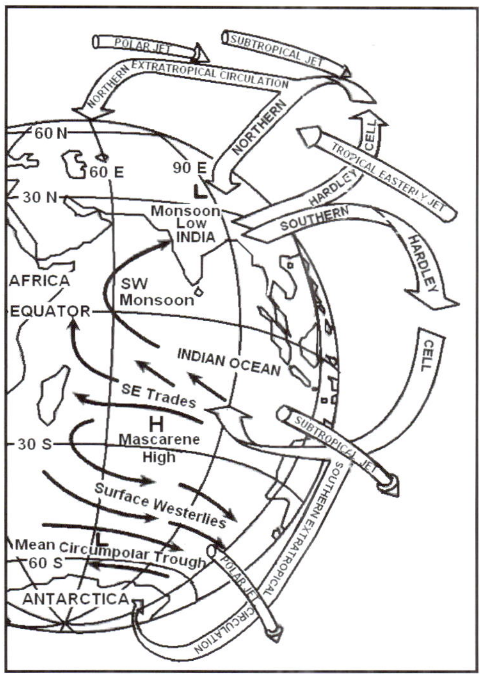

Fig. 7.1.2. Schematic view of meridional circulation in the northern summer, reproduced from Meehl (1987). The rising air from tropical heating moves either north or south, and sweeps downward over subtropical latitudes in both hemispheres. Reverse circulation occurs where polar air that rises in high latitudes joins the downward branch of the Hadley cell.

The air above the ground is hot and unstable in the tropics. Afternoon showers are common. When upper-level troughs approach or cold air masses move southward, thunder clouds intensify. Current computer models have difficulty resolving convective clouds in the tropics. Conceptual models based on mesoscale model products will be discussed in Chapter 9 to aid the prediction of potential thunderstorm development.

Cloud activity is persistent over the ITCZ or Near-Equator Trade Wind Convergence (NETWC), where trade winds in both hemispheres merge. In the northern summer, the ITCZ shifts a few degrees northward from its normal position near the equator. Another distinctive feature near the southern tropics is the convergence zone over the northern part of South America that connects with the heat low extending to western Colombia, the Gulf of Panama, and Bolivia. Extensive rainfall occurs in those regions during northern summer, and often intensifies over the high mountain ridges, which are known the South Pacific Convergence Zone (SPCZ), a reverse-oriented monsoon trough.

Generally speaking, weather regimes rooted in low latitudes are more difficult to predict with computer models than those in high latitudes. This is mainly due to the limitation in analyzing moisture fields and in simulating moist physical processes. Human experience and expertise to correct the model bias are essential to help improve computer-generated charts, particularly for precipitation regions.

7.2 Large-scale land-sea breeze

The ideal picture of zonal mean circulation is more complicated when geographical variation is considered along a circle of latitude from east to west. The land and ocean have different thermal capacities with respect to solar heating, from which large-scale sea breeze circulation is derived. The continent is warmer than the adjacent ocean in the summer hemisphere. Warm air that rises over the heated continent subsides onto the adjacent ocean, establishing a giant whirl. The surface wind blows toward the warm continent, and in return, upper-level winds blow toward the cold ocean. Stationary highs and lows correspond to the downward and upward branches of the circulation respectively along the latitude circle in Fig. 7.2.1.

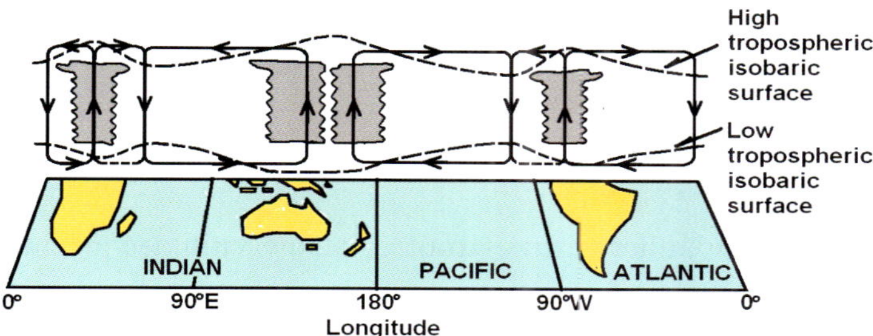

Fig. 7.2.1. Schematic view of east-west atmospheric circulation along the longitude-height plane over the equator. The figure is reproduced from Webster (1983). In the longitude-height cross-section, the rising motion occurs over Indonesia, and the compensating downward motions are found over the East Pacific and West Indian Oceans.

The time-mean positions of permanent highs and lows are presented in Fig. 7.2.2 for winter and summer, respectively, where the daily fluctuation of pressure is smoothed out. The dominant feature in the pressure map is the maritime high and continental low in the summer hemisphere, and the maritime low and continental high in the winter hemisphere. That feature is more evident in the northern hemisphere where land-sea contrasts are prominent. The Pacific high and Bermuda high expand over the ocean and the thermal low over the Asian-African continent and North America extends to the high latitudes in the northern summer. On the other hand, the Siberian high and Canadian high become dominant, and the Aleutian low and Icelandic low become stronger during the northern winter. Maritime highs in the southern hemisphere become stronger in summer, but their positions change little except for the Australian region.

The wind patterns change with the seasons. Cold and heavy air masses build up in the center of major continents in high latitudes, and migrate to the central and eastern parts of the major continents in winter. At times, cold surges occur as artic cold air masses abruptly march equatorward with the expanding continental high pressure system. A prominent example of continental climate is found over northeast Asia and the northeastern seaboard of the United States. On the other hand, mild and humid conditions are prevalent on the western part of continents in high latitudes. The permanent low over the ocean to the west and highs over the continent to the right give rise to advection of warm air from low latitudes over these regions.

The situation is reversed in summer. The eastern part of the continent receives hotter and more humid air from low latitudes, and severe convective systems often develop in the region. On the other hand, stable and dry conditions with occasional fog prevail in the western part of the continent. The upwelling of deep cold water further strengthens stability in the vicinity of the oceanic high. The western coasts of America, southwestern Europe, and South Africa are good examples.

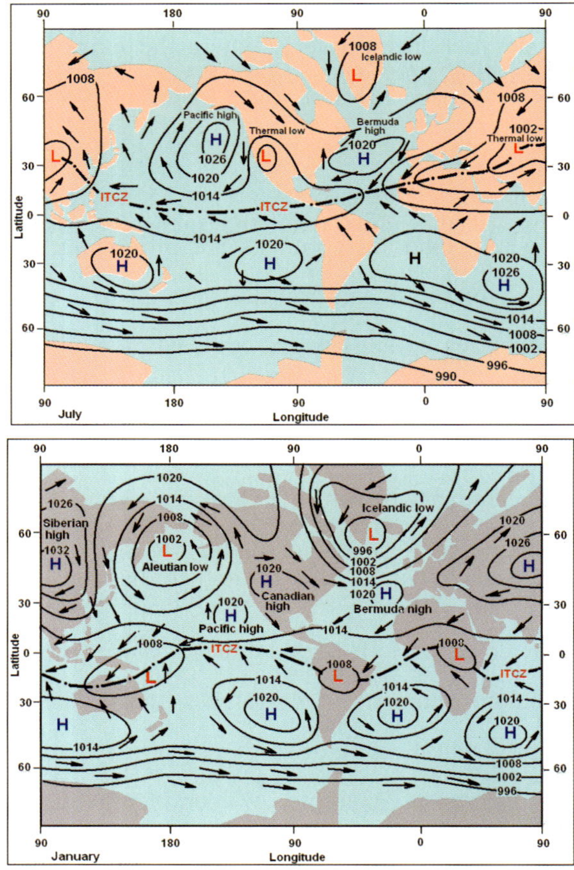

Fig. 7.2.2. Mean sea-level pressure distribution and surface wind-flow patterns for July (upper) and

January (lower). The heavy dashed line represents the position of the Intertropical Convergence Zone (ITCZ) or in other words, the Near-Equator Trade Wind Convergence (NETWC). In the summer hemisphere the ocean is cooler than the continents. The same mechanism as that underlying land-sea breezes supports the high pressure system over North Pacific and North Atlantic in July. In contrast, the continent is relatively colder than the ocean in the winter hemisphere, and the Siberian high and Canadian high are dominant in January. Seasonal variation of pressure systems is weak in the southern hemisphere, as the proportion of land cover is small. The figure is reproduced from www.unca.edu/~dmiller/Chapter%207.ppt

The Asian continent has the largest land area and the highest peak in the world. Its seasonal wind has unique features commonly known as the Asian monsoon. The massive heat source over the Tibetan Plateau and surrounding deserts drive the monsoon circulation, which is a major element representing seasonal change that covers wide areas centered over South Asia. Its influence extends from the African coast to the Eastern Pacific, and from Australia to the north of the Tibetan Plateau.

Seasonal winds from maritime air masses drives warm moisture toward the monsoon trough. Most of the rains over the monsoon area fall in summer or autumn with the southwesterly monsoon. However, equatorial regions such as Singapore, Taiwan, and the Philippines also record maximum rainfall in the cold season with the northeasterly monsoon. The monsoon regions with double rainy seasons are hatched in Fig. 7.2.3.

The southwesterly monsoon flow during the northern summer starts with the trade winds in the southern hemisphere, crossing the equator near the sea of Madagascar and moves clockwise to pass through India, the Indochinese peninsula, and far north to the Korean Peninsula as seen in Fig. 7.1.2. Conversely, the northeastern monsoon that originates in cold Siberia brings easterlies to China, the Indochinese peninsula, India, Indonesia, and the Philippines during the northern winter.

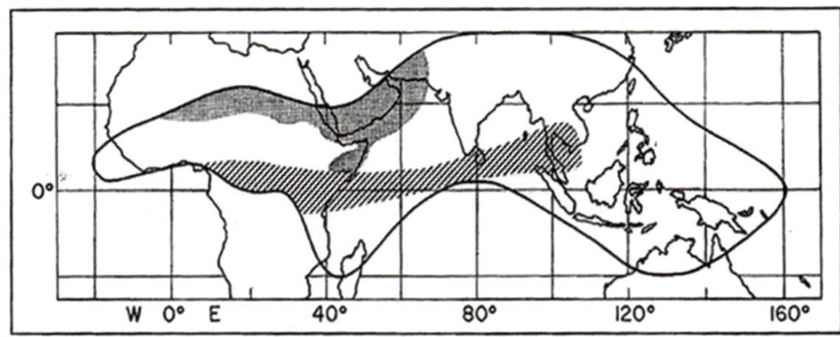

Fig. 7.2.3. Schematic diagram for the monsoon region adopted from Ramage (1995). The deserts, which have an average annual rainfall of less than 250 mm, are shaded in the enclosed area. Areas with a bimodal pattern of annual rainfall are hatched.

The evolution of weather systems is closely related to changes in planetary-scale circulation, which shows variations at longer intraseasonal time scales. A typical example of intraseasonal fluctuation is the Madden Julian Oscillation (MJO), which refers to cloud clusters near the equator propagating eastward over a period of 40~50 days. It affects not only the local weather with heavy rain and strong wind, but also in regions a few thousand kilometers downstream through Rossby wave dispersion. Travelling over warm oceans, it also influences monsoon and tropical cyclones (Moncrieff et al., 2007).

The El Niño Southern Oscillation (ENSO) is a prime example of interannual oscillation phenomena. In normal years, rising motion over deep cloud clusters around the Indonesian region spreads out in the upper troposphere. Its eastern branch descends over the east Pacific, while the western branch sinks west of the Indian Ocean. However, in El Niño years the main convective activity moves to the central Pacific, with the corresponding atmospheric circulation inducing abnormal climate over the monsoon region. Favorable locations for tropical cloud clusters and tropical cyclones are affected by the variation of sea surface temperature modulated by ENSO.

7.3 Long-range forecasts

Just like the daily short-term forecast, long-range forecasting is also based on the forecast chart, but averaged over longer periods. Time-mean charts can be interpreted in the same as ordinary ones. Let's take an example of a monthly mean 500 hPa chart. When the monthly mean 500 hPa trough (ridge) is located over a region, the region would be colder (warmer) than the surrounding area for the month. The region located downstream of the monthly mean 500 hPa trough is favorable for the frequent passage of storms in the month, while fair weather conditions are expected to persist upstream. Seasonal mean or annual mean charts are interpreted in the same manner as the monthly mean chart.

Departures from normal conditions are indicators of degrees of abnormality. The 30-year mean is often used to define the climatological normal. 500 hPa departure fields relative to the climatology tells us the degree of abnormality in the given state. The positive (negative) anomaly in the 500 hPa departure of geopotential means it is warmer (colder) than normal in the lower troposphere. For instance, in Fig. 7.3.1, there was a pronounced positive anomaly in geopotential height at 500 hPa over the Mediterranean, northwest and south Africa, northeast Asia, eastern Canada and Brazil from June to August 2008, which corresponds to warmer climate during the season. On the other hand, Northern Europe and eastern part of North America were colder than normal with negative anomalies in geopotential height. In the northern midlatitudes, the wet climate over northern Europe and east Alaska matched well with southwesterlies that bring moisture from the subtropics in agreement with the conceptual model in Fig. 4.2.1. Such relationship is not applicable to the tropical regions.

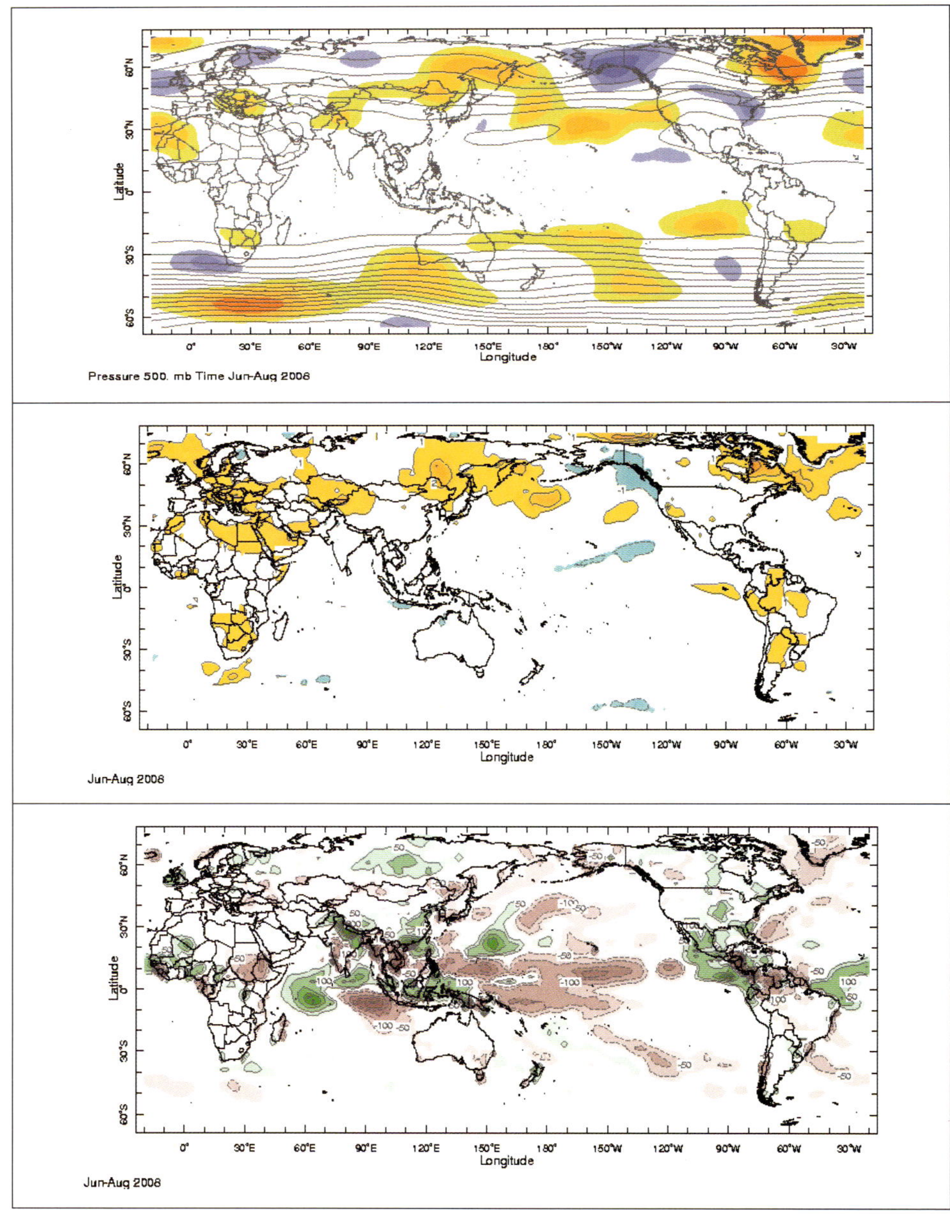

Fig. 7.3.1. NCEP seasonal mean climatology for June, July, August in 2008 in units of standard deviation. Geopotential height at 500 hPa with positive anomaly in yellow and negative anomaly in blue (upper), temperature at 2 m above ground with positive anomaly in yellow and negative anomaly in blue (Middle), and precipitation amount with positive anomaly in green and negative anomaly in purple (lower). Contour interval is CI= 0.4, 1 degree, and 50m/day respectively from upper to lower panels. The images are available from http://iridl.ldeo.columbia.edu/maproom/. Global/. Source: IRI.

Forecasting slow-varying fields

The weather pattern for the coming month or season is intimately connected with boundary conditions like sea surface temperature, sea ice, vegetation, and snow cover far from the target region. There are numerous studies reporting on the long-term fluctuation of temperature and precipitation at one place in relation to the surface boundary condition at another place. Such remote relationship is called a teleconnection pattern, a topic that scientists have been working on to improve long-range forecasting. The same inputs lead to different forecasts if different theories of teleconnection are applied. An example of a teleconnection pattern is shown in Fig. 7.3.2 for geopotential at 500 hPa with the variation of the El Niño Southern Oscillation index over the Eastern Pacific Ocean. The U.S. Climate Prediction Center (CPC) offers online the analyses and projections for the various teleconnection patterns including the Arctic Oscillation (AO), the North Atlantic Oscillation (NAO), the Pacific-North American Pattern (PNA), and the Antarctic Oscillation (AAO): http://www.cpc.ncep.noaa.gov/.

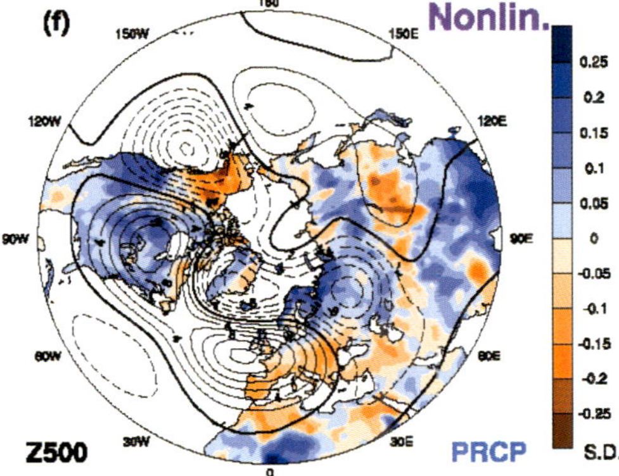

Fig. 7.3.2. An example of a teleconnection pattern for geopotential at 500 hPa (contour) with the variation of the El Niño Southern Oscillation index over the Eastern Pacific Ocean (Hsieh et al., 2006).

While those conceptual tools continue to gain popularity among forecasters, scientists rely more and more on dynamical prediction tools. Contemporary general circulation models (GCM) show promising results for prediction of planetary-scale motions on the seasonal and longer time scales, while medium-range forecast models extend the predictability of daily weather variation beyond 10 days with the advance of satellite data assimilation.

The seasonal forecast charts from APCC and ECMWF show pronounced positive anomaly in geopotental at 500 hPa over zonally elongated regions extending from Mongolia to the west coast of Canada, and the south coast of Australia to the coast of Chile in Fig 7.3.3. These features are well matched with the observed climate in Fig. 7.3.1. However, the forecast charts disagree on regional details. The pocket of positive anomaly over Western Europe is over-forecasted in the APCC forecast. The ECMWF forecast showed a slightly better match with the observed over Western Europe and eastern part of North America.

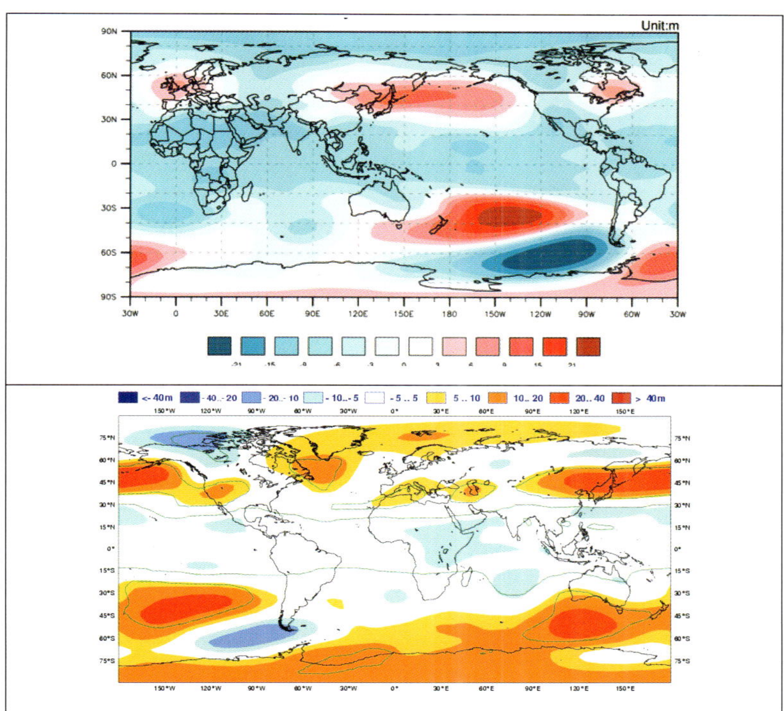

Fig. 7.3.3. Seasonal (June, July, August) forecasts from APCC (upper) and ECMWF (lower) for 500 hPa geopotential height anomaly in units of meter, starting from the initial condition for May 2008. The images are available from http://www.apcc21.org/en/services/forecasts/3mon/latest_monthly and http://www.ecmwf.int/products/forecasts/d/charts/seasonal/forecast/seasonal_range_forecast/ respectively. Similar charts can be found on the websites of the IRI, NCEP, and UKMet for precipitation and temperature.

Other websites for the updated long-range forecasts can be found in Table 7.3.1. Some site can be only accessed by registered users. Some websites such as the IRI and the APCC provide collections of model forecasts and their skills. CPC provides numerous forecast guidance charts associated with blocking, MJO, and monsoon activity. Figure 7.3.4 presents an example of MJO forecast from CPC.

Table 7.3.1. Websites for seasonal forecasts. For more information, visit the WMO directory at http://www.wmo.int/pages/prog/wcp/wcasp/clips/producers_forecasts.html.

Meteorological Center	URL (http://)
BoM (Australia)	www.bom.gov.au/climate/ahead/
CMC (Canada)	www.weatheroffice.gc.ca/saisons/index_e.html
CPTEC (Brazil)	www.cptec.inpe.br/clima/
ECMWF	www.ecmwf.int/products/forecasts/seasonal/
IRI (USA)	iri.columbia.edu/climate/forecast/
KMA (RoK)	web.kma.go.kr/eng/weather/forecast/long-range1.jsp
BCC (China)	bcc.cma.gov.cn/en/
TCC (Japan)	ds.data.jma.go.jp/gmd/tcc/tcc/index.html
CPC (USA)	www.cpc.ncep.noaa.gov/
UKMet (UK)	www.metoffice.gov.uk/research/seasonal/

Long-range forecast skill varies from region to region. In spite of the partial success in predicting seasonal climate associated with ENSO, simulations of the Asian monsoon and MJO are particularly challenging tests for climate models. Atmospheric sensitivity to ENSO is the most evident near the tropics, but weak in some midlatitude countries such as Korea during summer. The poor predictability for the Asian monsoon system severely limits the skills of long-range forecasts. Ensemble techniques are widely used to quantify forecast uncertainty, as effectively demonstrated on the APCC website.

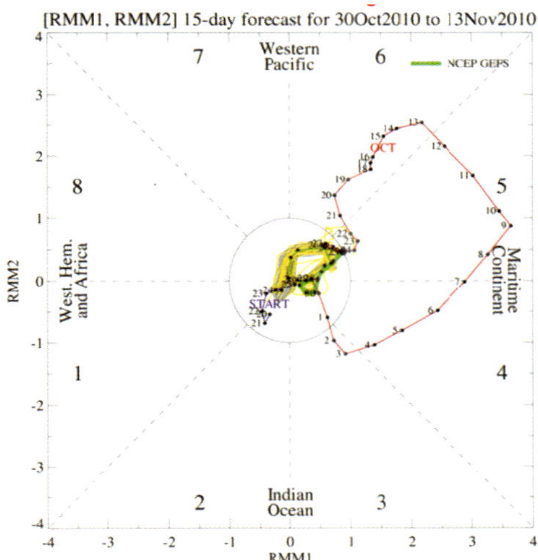

Fig. 7.3.4. Ensemble phase diagram for the Madden Julian Oscillation (MJO) index from the Global Forecast System (GFS), NCEP, showing the evolution of the last 40 days of observations along with the 15-day ensemble forecast. The yellow lines are the twenty ensemble members and the green line is the ensemble mean (thick-week 1, thin-week 2). The dark gray shading indicates that 90 % of the members fall in this area and the light gray shading indicates that 50% of the members do. The MJO index is presented in the RMM1 and RMM2 axes that represent MJO location and the intensity. MJO is weak and difficult to diagnose if its index is within the center circle. Usually the MJO moves counterclockwise in the phase diagram, and its location is identified by one of 8 phases in the diagram. The image is available from http://www.cpc.ncep.noaa.gov/products/precip/CWlink/MJO/foregfs.shtml. Source: U.S. NOAA.

Climatological frequency

While the dynamic approach suffers from forecast uncertainty, particularly at the regional scale, the statistical approach has been extensively used for application, mainly for its simplicity and economy. The teleconnection pattern discussed earlier is only one example of the statistical methods. One may even derive a simple formula for a long-range forecast from a time series of observation records. Let us take an example of the climatological data available from the KMA website. Mr. Kang wants to get the weather forecast 45 days ahead for his wedding in Seoul. The long-range forecast only tells him the monthly mean temperature and rainfall for the coming month, which is too crude to foretell the weather on the day of his wedding. Instead he can search for the past-years records for Seoul for the last 30 years on the KMA website (http://www.kma.go.kr/weather/climate/average_30 years.jsp). He found that there will be a 30% chance of rain on that day based on the climate records of the last 30 years. The information may be used as a basic guideline for his personal decision-making, when no weather forecast is available beyond two weeks in advance. A similar approach may be undertaken to predict the occurrence of other weather elements such as cold spells, cloudiness, snow, and fog. The WWIS website of WMO also provides the monthly climatology data including temperature, total precipitation, and number of precipitation days for the given city. More sophisticated users can apply a statistical downscaling technique to translate coarse resolution forecasts into fine resolution details on both temporal and spatial scales (Kang et al., 2007; Nguyan et al., 2002)

7.4 Toward climate change

These days, most of the advanced meteorological centers use climate models to project future climate change driven by both natural and anthropogenic forcings. The anthropogenic production of greenhouse gases such as CO_2, NO_2, O_3, which is mostly due to the combustion of fossil fuels, has increased rapidly in the last century. Increase of CO_2 and other green house gases would result in arise of 4 degrees in the global mean temperature by 2100, based on the International Panel on Climate Change (IPCC) report, other conditions being equal. The projection of green house warming is as uncertain as the projection of CO_2 concentration in the atmosphere, with the caveat that climate models differ considerably in simulating future and past climate.

The more important issue is climate change on the local scale. The projected surface temperature change in Fig. 7.4.1 for the late 21st century from the IPCC report reveals large differences from region to region. Would the tropical cyclones and midlatitude storms intensify? Would they threaten coastal communities more frequently? How about the maximum and minimum temperature? Would it change toward a more tropical or polar climate? What is the impact on the ecological environment, and on industries and businesses? These questions can be addressed by the climate models, which are designed to simulate the various sub-components of earth and their feedbacks on the atmosphere. Various components comprise the earth system, including chemical processes, biosphere processes near the land surface, ocean circulation, and polar ice processes. The current computing power cannot support sophisticated climate models with full physics that run on a high resolution less than 10 km. One exception is Japan's earth simulator, whose resolution reaches down to 5 km with 96 vertical levels (Ohfuchi et al., 2007).

Alternatively, downscaling experiments with nested models provide partial clues on local climate change. They use the boundary conditions supplied by coarse mesh models. The ensembles of the experiments are expected to filter out the random noise associated with the downscaling, and to provide a regional estimate of uncertainty on future scenarios. Various research groups are actively working on this problem, and their findings can be found in the reports of the IPCC (http://www.ipcc.ch) and associated websites.

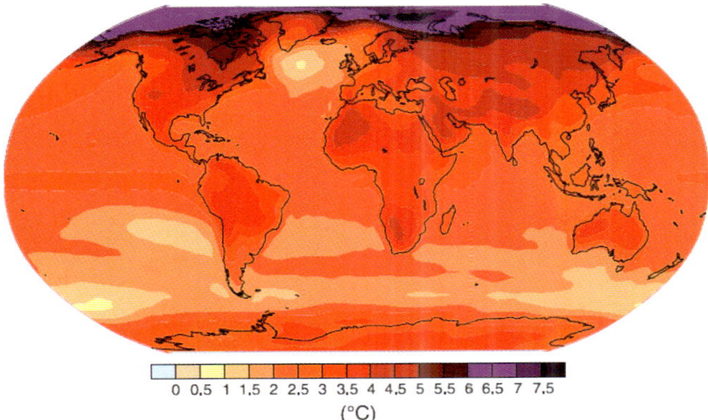

Fig. 7.4.1. Projected surface temperature change for the late 21st century from the multi-model average

for the A1B scenario, which assumes carbon dioxide emissions increasing until around 2050 and then decreasing after that year, as presented by the IPCC (2007). Temperatures are relative to the period 1980~1999.

8. Cyclone Forecasts

As a natural extension of the previous chapter, two large-scale atmospheric motions are considered in this chapter: tropical and extratropical cyclones, which are major ingredients for weather forecasting and climate prediction. This chapter briefly reviews the model characteristics of their simulated evolutions.

8.1 Tropical cyclones

Tropical cyclones that develop over the warm Pacific or Indian Oceans near the equator occasionally visit monsoon regions. They tend to move westward because of trade winds in the subtropics. Some may reach mid-latitude and track eastward following the polar jet stream. A cloud picture of tropical cyclones is provided in Fig. 8.1.1 as an example. Three tropical cyclones develop over the monsoon trough, accompanied by clusters of convective clouds along the cross-equatorial flow 10 degrees north of the equator. The International Best Track Archive for Climate Stewardship (IBTrACs) project archives best track data for tropical cyclones from as early as 1842, providing the distribution, frequency and intensity of tropical cyclones worldwide.

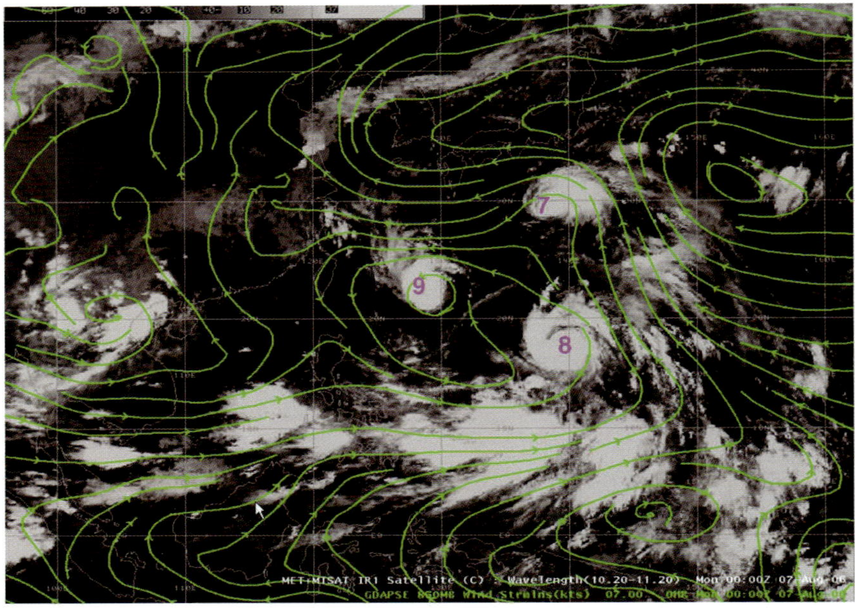

Fig. 8.1.1. Three typhoons (No.7 to 9) appear on MTSAT-1R Infrared image at 00 UTC on 7 August 2006. The streamlines at 850 hPa are presented in solid lines. Tropical wave activities are strong along the monsoon trough. Source: KMA.

A tropical cyclone is a multi-scale system. The eye-wall cloud with deep convection is only a few kilometers wide, while the outer flow boundary extends to several hundreds of kilometers from its center. The resolution of contemporary models fail to capture the mesoscale details of precipitation band structures in tropical cyclones. Furthermore, observation over the ocean mostly relies on satellite measurements, and large uncertainty exists in the estimation of the center position, rainfall rate, speed and radius of the maximum wind. Table 8.1.1 summarizes tropical cyclone position errors from aircraft-satellite comparisons (ASC), and satellite-satellite comparisons (SSC). The averaged position fix errors range from 45 to 55 km, varying with analysis techniques. Table 8.1.2 shows tropical cyclone intensity errors from the satellite consensus and the subjective Dvorak technique during 1999~2009. The averaged errors of intensity estimation amount to 5.2 hPa and 7.6 hPa for the satellite consensus and the Dvorak technique respectively.

Table 8.1.1 Tropical cyclone position errors in units of km from aircraft-satellite comparisons (ASC), and satellite-satellite comparisons (SSC). Details can be found in Martin and Gray (1993).

Difference statistic	ASC	SSC
Mean	44 km	56 km
Standard deviation	43 km	56 km
10th percentile	11 km	11 km
90th percentile	93 km	115 km
10th percentile to 90th percentile range	82 km	104 km
Number of cases	4588	2906

Some numerical prediction centers deliberately filter out the tropical cyclone vortex in the analysis, and fill up the gap using the bogused vortex reconstructed from the observed parameters of the tropical cyclone. On the other hand, satellite radiance data assimilation without the bogused vortex is so far successful and outperforms assimilation with the bogused vortex for medium-range tropical cyclone tracks, as demonstrated by the ECMWF forecast (EMX) in Fig. 8.1.2. Understanding the specified tropical cyclone vortex in the initial state facilitates conjectures on subsequent model errors.

Table 8.1.2. Tropical cyclone intensity errors for 460 cases during 1999~2009 from the satellite consensus (SATCON) and the subjective Dvorak technique, measured by accuracy of minimum sea level pressure (MSLP) in units of hPa and maximum sustained wind (MSW) in units of km/hr and knots (Morinari and Fukada, 2010). Dvorak is average of TAFB and SAB. MSLP and MSW are verified with reconnaissance flight measurements (Herndon et al., 2010).

	SATCON MSLP (hPa)	Dvorak MSLP (hPa)	SATCON MSW (km/hr, knots)	Dvorak MSW (km/hr, knots)
Bias	0.3	-2.7	-1.9 / -1.0	-5.6/ -3.0
Average error	5.2	7.6	13.3/ 7.2	15.0/ 8.1
RMS error	6.4	9.1	15.4/ 8.3	16.7/ 9.0

Model Consensus

The consensus of global model forecasts available from major operational centers, a kind of EPS, is widely used for tropical cyclone forecasting. The simple consensus has proved to be more accurate than any individual model forecasts. More sophisticated EPS with perturbed initial states and stochastic forcings, as discussed in Chapter 6, can be applicable to tropical cyclone forecasting. One should be on the alert, however, for any systematic bias in the model forecasts. In the extreme case, all models can fail. Elsberry and Carr (2000) demonstrate that about 8% of cyclones fall into such cases.

Limitation of model forecast

Averaged tropical cyclone track forecast error from state-of-the-art models reaches 200 km for D+3 and 310 km for D+5, as shown in Fig. 8.1.2. The obvious defect in the model forecast comes from the imperfect initial condition for the tropical cyclone vortex and its environment. Microwave sensors can detect the precipitating particles below the cloud shield. Assimilation of these radiance data makes it possible to incorporate the inhomogeneous features of precipitation bands into the initial analysis. Up-to-date satellite measurements are not sufficient to capture the details of 3-dimensional structure of a tropical cyclone. The model track and intensity forecasts derived from the incomplete initial conditions often deviate considerably from the observed, particularly in the case of weak storms, and are subject to adjustment during the forecasting process. Figure 8.1.3 shows the consensus model D+1 track forecast error versus initial tropical cyclone intensity for the Atlantic basin during 2001 to 2003 (Goerss, 2007). The consensus model D+1 track forecast error ranges from 121 km (75 nautical miles) for weak tropical cyclones with an initial intensity of 56 km/hr (30 kt), to 40 km (25 nautical miles) for strong tropical cyclones with an initial intensity of 241 km/hr

(130 kt). The model track can be simply corrected by subtracting the track bias from the model track forecasts throughout the forecast period. Such linear extrapolation is no more effective beyond a day or two at which the nonlinear effect becomes dominant.

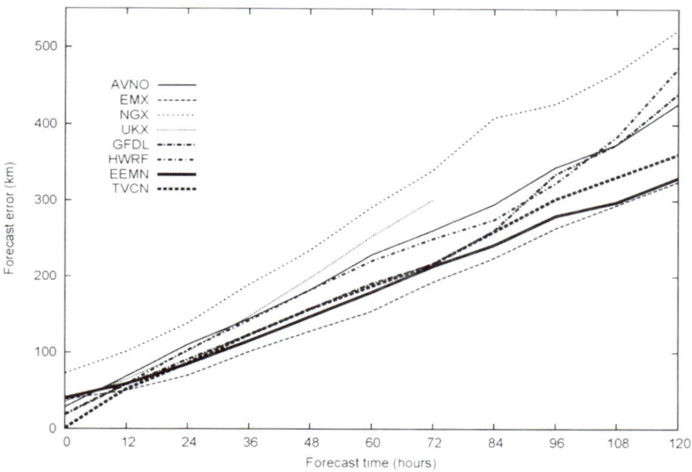

Fig. 8.1.2. Averaged tropical cyclone track forecast errors of various global models in units of km for 159 Atlantic tropical cyclone cases. Details can be found in Majumdar and Finocchio (2010).

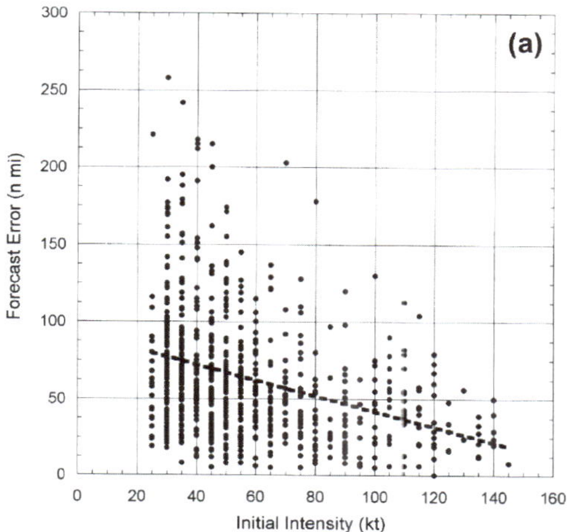

Fig. 8.1.3. Consensus model D+1 track forecast error in units of nautical miles versus initial tropical cyclone intensity in units of knots for the Atlantic basin during 2001~ 2003 (Goerss, 2007).

Forecasting tropical cyclone intensity is a challenging subject in the modeling community. Even the resolution of high-resolution models is insufficient to represent small-scale features near the center of a tropical cyclone. The model physics do not cover all the details of nonlinear interactions across the wide range of scales of motions embedded in a tropical cyclone. Figure 8.1.4 shows the intensity error cumulative distribution of the U.S. National Hurricane Center's official forecast for the Atlantic basin tropical cyclones during 2005~2009. More than 70% of the official D+2 intensity forecasts have errors over 37 km/hr (20 kt) in terms of maximum wind speed, which increase proportionally with lead time.

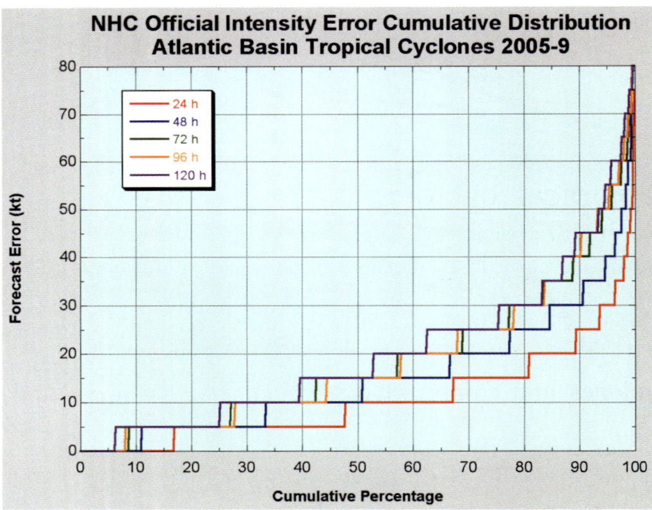

Fig. 8.1.4. Intensity error cumulative distribution of the U.S. National Hurricane Center's official forecast for Atlantic basin tropical cyclones during 2005~2009. http://www.nhc.noaa.gov/verification/verify4.shtml. Source: U.S. NOAA.

Forecasting of tropical cyclone formation is an emerging area. The state-of-the-art model can predict its formation for D+1 and D+2 with accuracy over 60%, as shown in Table 8.1.3. A high-resolution limited-area model coupled with an ocean model is promising, but has yet to prove its utility. In addition, a limited-area model is influenced by the boundary condition supplied from a coarse mesh model that is never perfect. Model performance has to be carefully monitored before any extensive application. For the time being, information from statistical models could be as valuable as dynamical models for tropical cyclone intensity.

Table 8.1.3. Tropical cyclone formation hit rate +/- 1 day and +/-3 degrees from the high-resolution version (T799) and low-resolution version (T399) of ECMWF model during July 2008 ~ January 2010 (Harr, 2010).

	D+1	D+2	D+3	D+4	D+5	D+6	D+7
High-resolution (T799)	72%	66%	57%	5[illegible]%	41%	31%	28%
Low-resolution (T399)	66%	61%	49%	42%	40%	32%	33%

8.2 Extratropical cyclones

Weather forecast for high latitudes is primarily determined by the position and intensity of extratropical cyclones. Jet stream crganizes baroclinic zones that extratropical cyclones and fronts travel through. The position and strength of the jet stream vary not only on the short-term but also on the long-term time scale. It is always necessary to monitor the position of the jet stream and its evolution to look out for the storm's path.

Figure 8.2.1 shows track guidance from U.S. CPC, based on the tracking of low-pressure centers in the sea-level pressure field. Most extratropical storms travel along the latitudinal belt at 30°~70° in either hemisphere. They move closer to the equator in the winter hemisphere, and closer to the poles in the summer hemisphere. They tend to avoid high mountain areas and to deviate either to north or south, as seen downstream of the Rockies and the Tibetan Plateau in the northern winter.

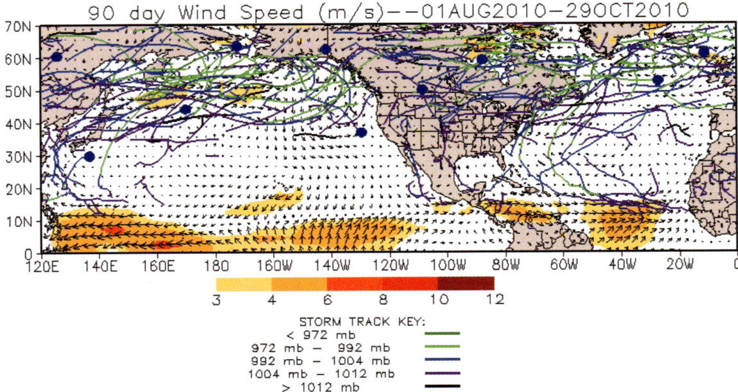

Fig. 8.2.1 Storm track guidance from U.S. CPC with varying intensity (color lines), and mean wind speed in units of m/s during the last 90 days. Storm tracks are identified by locating grid points in which the sea-level pressure is less than its surrounding grid points by at least 1 hPa. The storms are tracked by analyzing the position of systems between time steps and applying a maximum distance threshold between candidate pairings (800 km) and additional quality control checks. Red/blue dots indicate active storms as of 18 UTC for the respective plot ending date. The image is available from http://www.cpc.ncep.noaa.gov/products/precip/CWlink/stormtracks/mstrack.shtml. Source: U.S. NOAA.

Computer models are pretty good at predicting paths and intensities of extratropical cyclones and their accompanying frontal structures. Applying the concepts in Figs. 4.2.1 and 4.3.1, the weather forecasts along the storm tracks can be derived from the computer-generated extratropical cyclone maps. The rest of this section deals with the characteristics of extratropical cyclone forecast errors.

Numerical Errors

Truncation error in a numerical solution of a hydrodynamic equation is unavoidable as long as model grid spacing is finite. The case of a controlled numerical experiment is illustrated in Table 8.2.1. Magnitude of an ideal wave tends to attenuate, and its phase lags behind the analytic solution during the advection process.

Table 8.2.1. Characteristic of numerical errors for advection schemes in terms of damping rate per time step, and relative phase per wavelength (Riddaway, 2002). △x: unit grid distance.

(a) Damping/ time step					
	2△x	3△x	4△x	6△x	10△x
Upstream differencing	0.00	0.50	0.71	0.87	0.95
Crank-Nicholson	1.00	1.00	1.00	1.00	1.00
Lax-Wendroff	0.50	0.76	0.90	0.98	1.00
Gadd	0.13	0.79	0.95	0.99	1.00
Leapfrog	1.00	1.00	1.00	1.00	1.00
4th-order leapfrog	1.00	1.00	1.00	1.00	1.00
Spectral	1.00	1.00	1.00	1.00	1.00
Finite element	1.00	1.00	1.00	1.00	1.00
Semi-Lagrangian (cubic spline)	0.00	0.88	0.97	1.00	1.00
(b) Relative phase error					
	2△x	3△x	4△x	6△x	10△x
Upstream differencing	1.00	1.00	1.00	1.00	1.00
Crank-Nicholson	0.00	0.41	0.63	0.81	0.93
Lax-Wendroff	0.00	0.58	0.75	0.88	0.95
Gadd	0.00	0.98	1.03	1.04	1.02
Leapfrog	0.00	0.43	0.67	0.86	0.95
4th-order leapfrog	0.00	0.65	0.89	0.99	1.00
Spectral	1.05	1.02	1.01	1.00	1.00
Finite element	0.00	0.87	0.99	1.01	1.00
Semi-Lagrangian (cubic spline)	1.00	1.00	1.00	1.00	1.00

Numerical error characteristics can be also confirmed in sophisticated weather prediction models. The propagation speed bias for the model extratropical cyclones is summarized in Fig. 8.2.2. The phases of synoptic disturbances tend to lag behind the observed by 20~60 km/day. On the other hand, the planetary-scale waves move at a very low speed or stay stationary, and have relatively little bias.

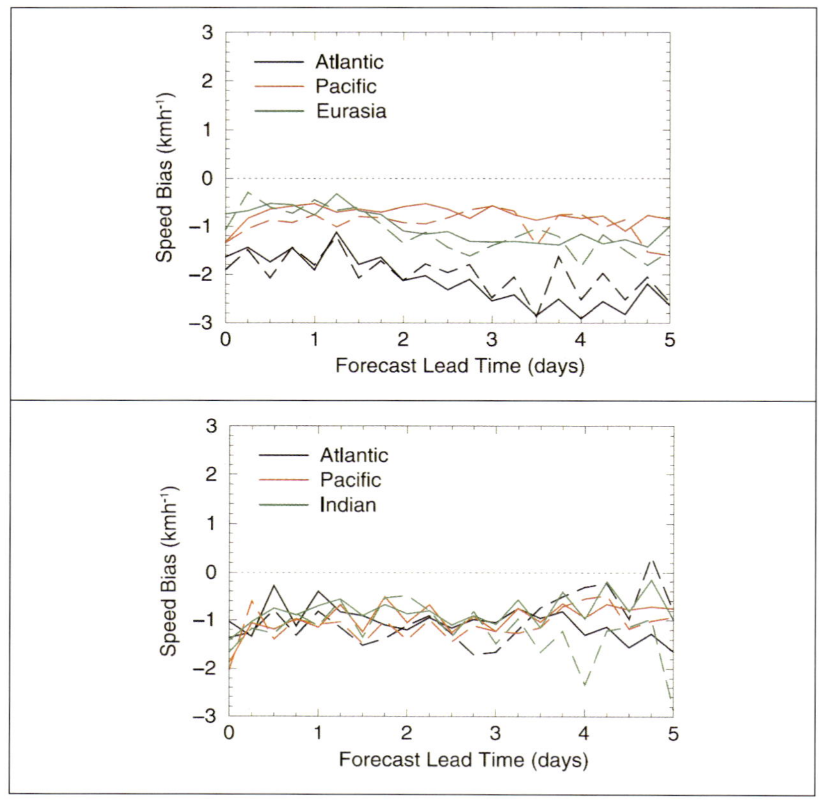

Fig. 8.2.2. Propagation speed bias in units of km/hr for extratropical cyclone vortices in the northern hemisphere (upper) and southern hemisphere (lower). The perturbed member tracks (dash line) and control forecast tracks (solid lines) are derived from the ECMWF ensemble forecast system (Froude, 2009).

Like the tropical cyclone vortex, the extratropical cyclone vortex is subject to numerical diffusion. Its amplitude tends to decrease by 0.2 $\times 10^{-5} s^{-1}$ per day, as shown in Fig. 8.2.3. The vortex, however, can grow through hydrodynamic instability process and other physical forcing mechanism in the model, which may offset the effect of numerical diffusion. Thus it is recommended to explore the physical reasoning behind the growth or decay of model disturbances prior to any adjustment.

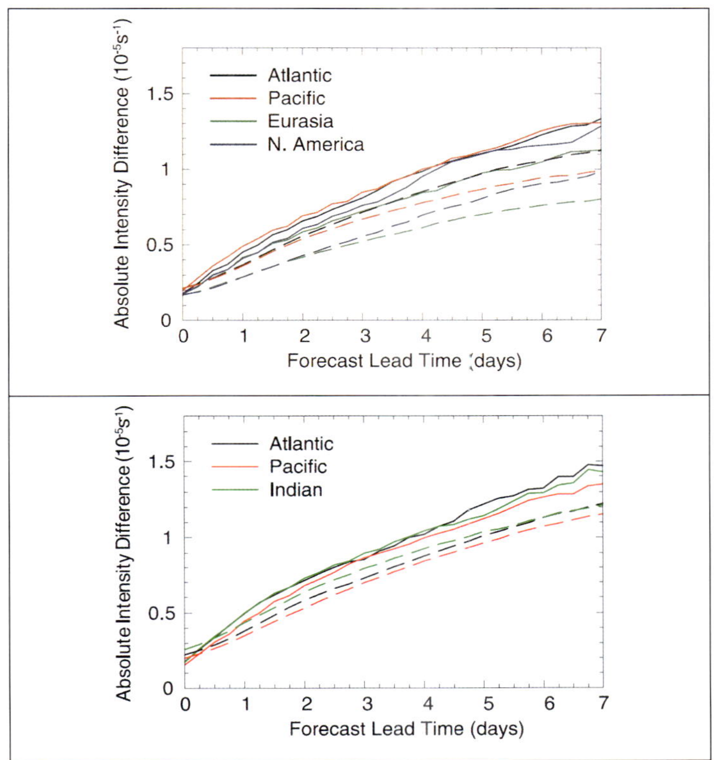

Fig. 8.2.3. Intensity error in units of $10^{-5}s^{-1}$ for extratropical cyclone vortices in the northern hemisphere (upper) and southern hemisphere (lower). The perturbed member tracks (dash line) and control forecast tracks (solid lines) are derived from the ECMWF ensemble forecast system (Froude, 2009).

Limitation of observations

Models generally follow the principle of large-scale dynamics fairly well, and users rarely need to modify the prognostic flow patterns, particularly for D+3 and onwards. However, there are some areas where users have to pay attention to see if the model goes in the right direction.

First, growth of an extratropical cyclone and its movement depend on latent heat release at the expense of available moisture. Storms over the ocean can grow more vigorously than those over land. If the storm in a model takes in less moisture than the observed, its intensity and accompanying precipitation are underestimated. A storm coming from data-sparse area needs extra caution as moisture field uncertainty increases.

Second, underactive convective parameterization in the model often leaves extra moisture in the lower atmosphere, which is used for the production of excessive latent heat. The overdeveloped storm tends to propagate farther poleward at a slower eastward speed.

Part Ⅵ. Fine-scale Details

9. High-resolution Models

9.1 Spatial resolution

A typical weather system consists of a chain of subsystems with different spatial and temporal scales. Figure 9.1.1 illustrates relationship between the characteristic scale and lifetime of a typical weather system. Planetary waves and cyclone waves generate available potential energy to fuel small-scale systems such as fronts, rain bands, and cumulus convections. In turn, those small-scale systems once developed return part of their energy to the large-scale waves.

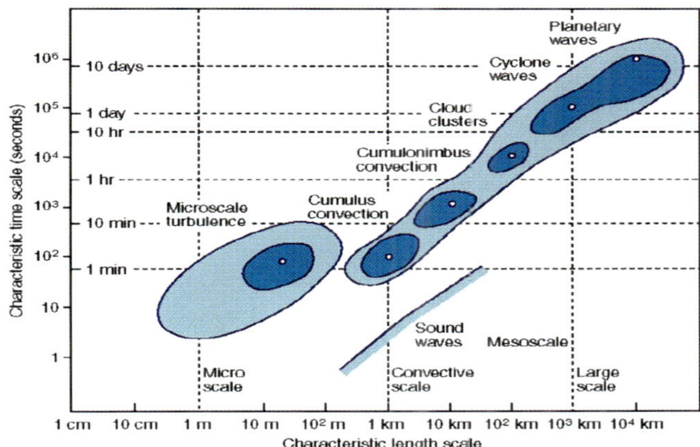

Fig. 9.1.1. Relationship between the characteristic scale and lifetime of a typical weather system in the atmosphere, taken from http://www.ecmwf.int/staff/roberto_buizza/Work/Lectures/Confs_WSs_Sems/IoP_CWP_18jan01.pdf. Predictability drops rapidly as the size of a weather system decreases. For instance, cumulus convections producing severe thunderstorms have time scales of only hundreds of seconds, while cyclones have time scales of days.

A model can practically resolve scales of motion greater than 5~10 grid units, as numerical errors easily contaminate the small-scale end of the wave spectrum. Furthermore, small-scale noise is often intentionally smoothed out during post-processing and visualization. The remaining small-scale features in the model fields should be examined with extra caution. A model with a grid spacing of 50 km can describe a extratropical cyclone with scales of 500 kilometers but not the embedded fronts, as illustrated in Fig. 9.1.2. The evolution of a block is not well simulated in a low-resolution model, as the model cannot properly explain multi-scale interaction and wave breaking of fine potential vorticity filaments. Generally, a low-resolution model overestimates the block's movement speed and tends to predict its decay to occur earlier than observed. Eye-wall clouds and convections embedded in a tropical cyclone are poorly represented in contemporary global models with a grid spacing of 20 km. The model with a grid spacing of 5 km can represent convective cloud clusters with scales of a hundred kilometers, but not the individual storm cell.

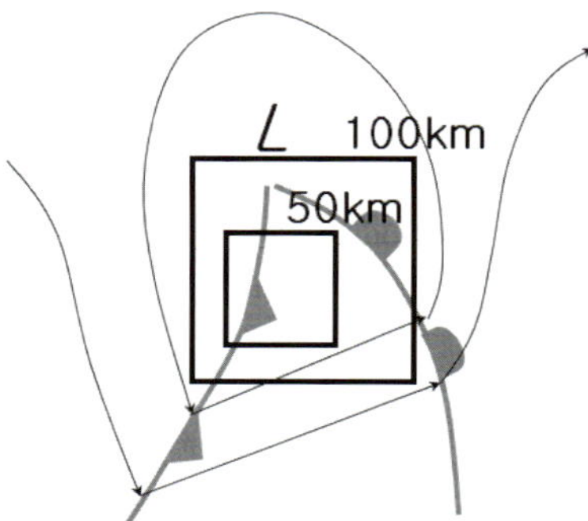

Fig. 9.1.2. Schematic relationship between model resolution and weather systems embedded in an extratropical cyclone. A model with a 50-km grid spacing can reasonably resolve a wave with wavelengths over 5 times of 2 unit grids, i.e., 500 km. Even though grid spacing is comparable to the width of a front, numerical errors cannot allow the model to predict waves shorter than cyclone-scale.

The timing and place of mesoscale weather events are difficult to predict even with a high-resolution model due to incomplete model physics and data assimilation. Those events are often associated with the weakly forced environment, mesoscale interactions including convective outflow boundary, and feedbacks among physical parameterizations such as PBL mixing and cloud radiative forcing. Model forecasts sensitive to microphysics parameterization include the formation of cold pool, size and density of a trailing stratiform area, and propagation of mesoscale systems (COMET, 2010). Maintaining consistency among the subsequent forecast runs is also a challenge. Model defects can be partially mitigated by introducing EPS, as discussed in Chapter 6.

There are two areas, however, where benefits of higher-resolution models can be expected; frontal and orographic precipitation. The essence of frontal dynamics is reasonably well covered in any model. Higher-resolution models reveal the finer structure on a front. However, the phase speed and strength of the front depend on the initial condition and synoptic-scale forcing during its evolution. Up-to-date radar images and rain gauge measurements are essential to adjust the model precipitation along the front.

Orographic precipitation depends on wind speed, height of the barrier perpendicular to the wind, vertical stability, and moisture distribution. As the model resolution increases, the wind flow crossing mountain barriers and forced condensation process can be described more precisely. The intensity of orographic precipitation in a model is inversely proportional to the grid spacing. This empirical relationship is useful for adjusting orographically forced precipitation in the model.

9.2 Non-hydrostatic models and thunderstorms

Some atmospheric phenomena such as thunderstorms are rarely seen in a synoptic analysis or forecast chart. This is partly because they are not well captured by the observation network, and partly because their characteristics are smoothed out during numerical analysis. Nonetheless, their existence strongly influences local weather and climate. In order to predict small-scale features, the model resolution should be enhanced accordingly. A high-resolution model with a

4-km grid spacing can resolve mesoscale elements such as rear inflow jets, surface outflow boundaries, convective mode as individual cells, and structures containing a leading convective line and trailing stratiform area (COMET, 2010).

Hydrostatic assumption is no longer valid for short waves that appear in high-resolution models. Most high-resolution models are based on non-hydrostatic equations that include vertical advection, acceleration of vertical momentum by buoyancy force, and dynamic pressure associated with air elasticity. Non-hydrostatic terms make it possible to simulate meso-vortices with tilting of vertical filaments, and lead to more reasonable updrafts with slantwise convections and gust fronts with downdrafts. Some useful parameters produced by high-resolution models are shown in Table 9.2.1. Updraft helicity, a product of vertical motion and vertical vorticity in the lower part of a storm, is related to storm rotation and tornado potential. Vertical integration of graupel and maximum updraft velocity indicates potential for hail. Maximum downdraft velocity is often used to examine potential occurrences of microbursts and Derechos. It also helps forecast dust storms. In dry climate, dust storms often progress over dry land with a gust front that forms along the leading edge of a cold pool associated with the downward branch of a thunderstorm. They often form, as cold downdrafts lift dust or sand into huge dark clouds which may cover over a hundred kilometers.

Table 9.2.1. Additional parameters useful for mesoscale forecast.

Target	Basic variables used for their derivation
Thunderstorm	Convective available potential energy (CAPE) Convective inhibition (CIN) Downdraft CAPE Storm relative wind
Severe thunderstorm	Vertical shear
Heavy precipitation	Simulated reflectivity (dBZ)
Gustness	Maximum 10 m wind speed
Hail	Maximum vertically integrated graupel

Target	Basic variables used for their derivation
Tornados or large hail	Updraft helicity
Large hail	Maximum updraft velocity
Microburst or Derechos	Maximum downdraft velocity

The simulation of mesoscale phenomena using high-resolution models is less than satisfactory for the time being, until model grid spacing reaches 1 km in scale and can resolve individual storm cell motions. Even if a model can reasonably describe the structure of the squall line and associated bow echo, their timing and intensity often considerably depart from reality. For instance, a model could reasonably predict the area of convective heavy rainfalls in the vicinity of an unstable subtropical high. Figure 9.2.2 shows a radar image of heavy rainfall (square box) along a weak frontal boundary on 11 July 2007. The 42-hr accumulated rainfall in the Weather and Research Forecast Model (WRF) from the previous morning run could only predict mild rainfall in the target region. Even if the model could begin to capture heavy rainfall from the previous night run and same-day morning run, the model would not be able to pinpoint the timing of the arrival of the line-shaped rainband in the blue circle. To be on the safe side, users are advised to take primarily the potential area of storm development from the model simulation, and refine details with reference to the latest observations and conceptual models. The convective parameterization scheme is not effective for models with a grid spacing below 5 km. Model precipitation without convective parameterization scheme is delayed until sufficient cloud particles and raindrops spin up. Users need to understand the spin-up problem, particularly early on in the projection period.

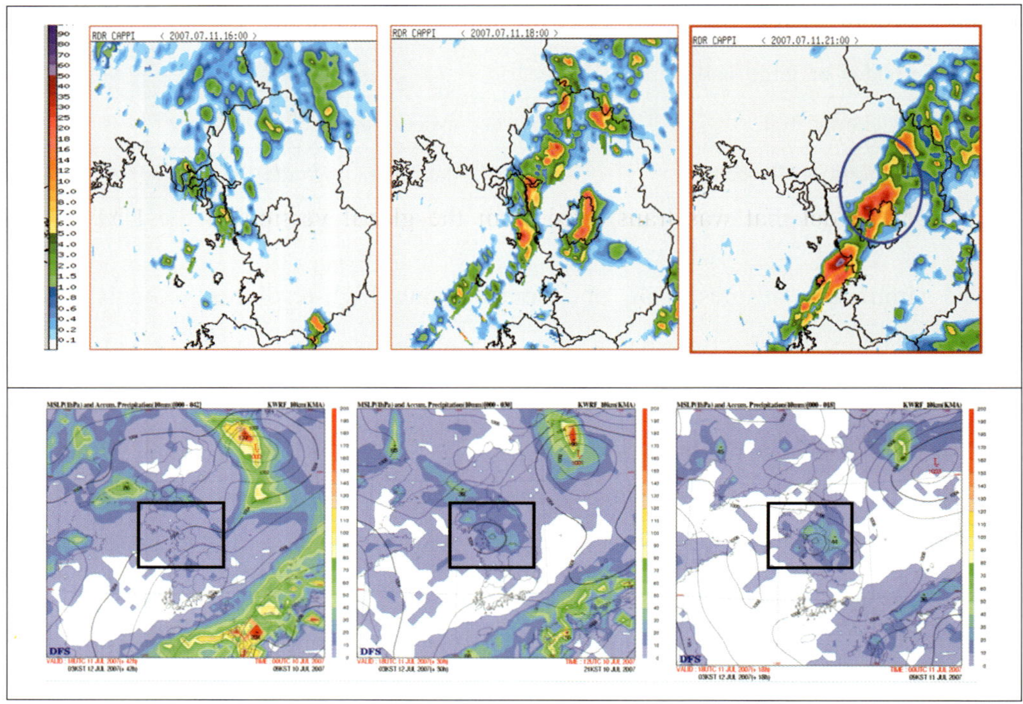

Fig. 9.2.2. Radar reflectivity at 15, 18, 21 LST on 11 July 2007 (upper), and accumulated precipitation from the WRF community mesoscale model with a 10-km grid spacing and 40 levels for durations of 42 hrs (lower left), 30 hrs (lower middle), and 18 hrs (lower right), terminating at 03 LST on 12 July 2007. The model starts from the initial state at 09 LST on 10 July 2007. The area covered by the radar reflectivity is highlighted by square boxes in the lower panels. Source: KMA.

Most high-resolution models are confined to the regional domain and require lateral boundary condition information from global models, which specify large-scale forcing along the boundaries. The smaller the domain, the greater the influence of the lateral boundary condition on a high-resolution limited area model. For example, forecast error in a steering flow from the global model is transferred to tropical cyclone track forecasts in the high-resolution limited area model, particularly for the weak tropical cyclones traveling along the Pacific Ridge. Figure 9.2.3 illustrates the systematic track bias of a high-resolution limited area model forced by a global model. The typhoon Kompasu was positioned east of Taiwan at 21 LST on 3 September, moving northward rapidly to reach the nose of Jeju

Island after 48 hours. However, both the global and regional versions of the UM model predicted a northwestward path for the typhoon at a slower speed of movement, and failed to correctly forecast the D+2 position of the typhoon. This is caused by the westward bias of the steering wind field in the regional version of the UM model that was transferred from the global version of the UM model.

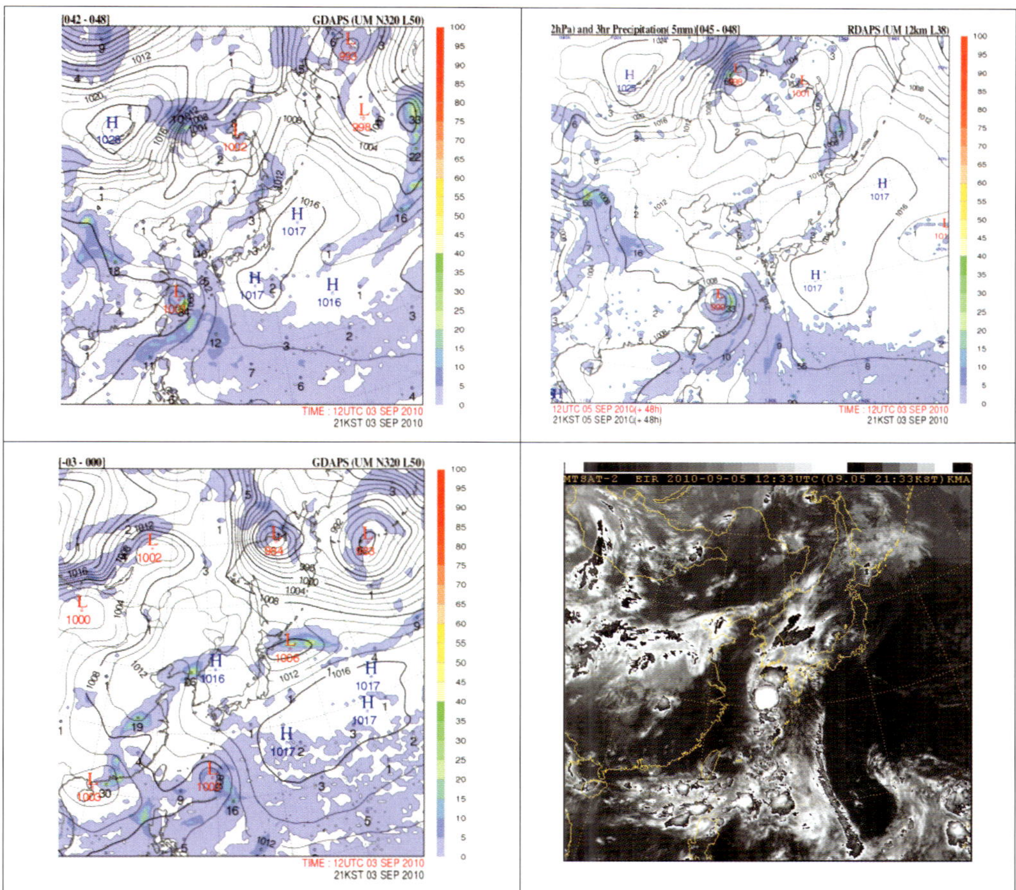

Fig. 9.2.3. Example of large track errors in a regional version of the UM model with a grid spacing of 12 km and 38 levels, given the lateral boundary condition from a global version of the UM model with a grid spacing of 40 km and 50 levels through one-way interaction. D+2 forecasts of mean sea-level pressure and 3-hr accumulated precipitation (color) from the global version (upper left) and the regional version (upper right), and enhanced infrared image (lower right) valid 21 LST on 5 September 2010. The initial state of mean sea-level pressure at 21 LST on 3 September 2010 is presented in the lower left panel. Source: KMA.

Theoretically, numerical errors at the lateral boundary propagate inside the limited area model domain at a speed of roughly a few tens of degrees per day. However, in operational practice it is hard to separate errors associated with the lateral boundary condition from other error sources without a thorough case study.

9.3 Predictability limit and nowcasting

Forecasting skill depends on the lead time. Figure 9.3.1 is a schematic of forecasting skills for different mesoscale short-range forecasts. Numerical weather prediction skill increases after the first 12 hours. A high-resolution storm model with radar data assimilation may help improve skill up to 3~12 hours ahead, but predictability decreases rapidly with time in the first few hours even with state-of-the-art technology. Simple linear extrapolation is effective for up to 1~3 hours; but its skill drops rapidly after that forecast range (National Research Council, 2010).

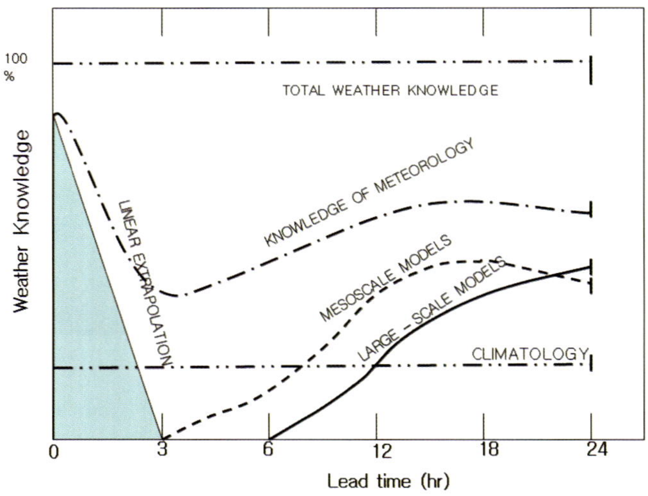

Fig. 9.3.1. Schematic of forecasting skills for different mesoscale short-range forecasts (Doswell, 1986).

Severe thunderstorms that bring high-impact weather conditions like lightning, hail, gusty winds, and heavy rain to local communities normally have a lifetime of less than 30 minutes in Fig. 9.1.1, while an organized group of storms or mesoscale convective systems may survive up to 6 hours or more. Their evolutions are very sensitive to initial conditions, and numerical models often fail to predict when and where a new cell will initiate due to the chaotic property of atmospheric systems. Favorable synoptic conditions provide the potential for storm development; but the specific timing and location depend on multi-scale interactions in the weather system. A storm-generated gust front often triggers a new storm cell downstream, which adds complexity to short-term storm forecasting.

An example is illustrated in Fig. 9.3.2 for the case of a severe convective storm at the mouth of the Hangang River, Korea, on 12 July 2001. In the early morning, an abrupt thunderstorm developed, followed by more than 150 mm of rainfall in just three hours near the metropolitan area of Seoul. This caused massive run-off and flash floods in small water channels as debris blocked the water streams. The sudden development of carrot-shaped storm clouds on the edge of the thermal boundary is well captured in the sequence of satellite images. The mesoscale model could predict only a few tens of millimeters of rainfall for the event far south of the actual rainfall area. It was very challenging to forecast the timing and location for the initiation of storm cells even for a few hours ahead.

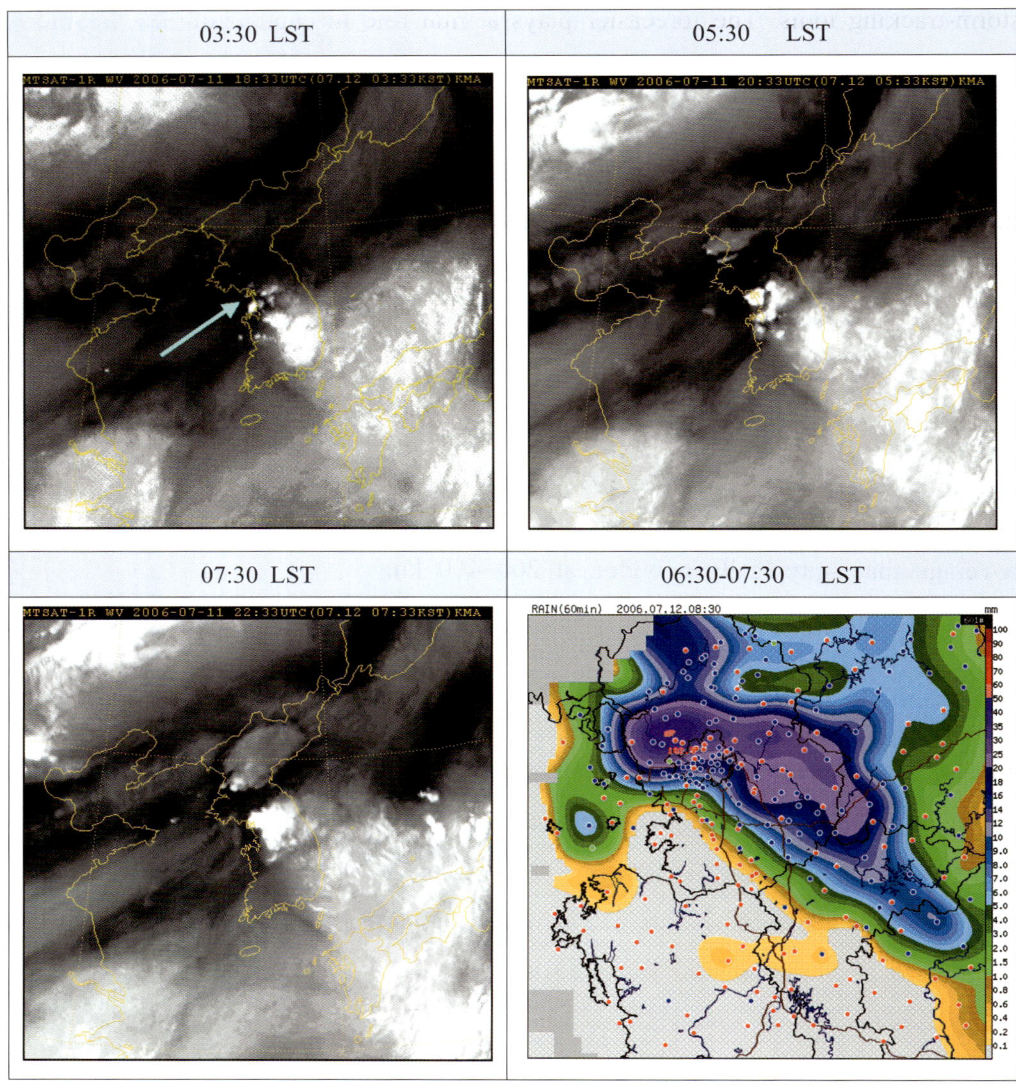

Fig. 9.3.2. An example of the genesis of severe thunderstorm clusters near the mouth of the Hangang River, Korea (marked by blue arrow). The heavy rains exceeded 50 mm per hour and persisted for several hours in the northern suburb of Seoul at dawn on 12 July 2006. The first three figures show the water vapor channel imagery from MTSAT-satellite at 03:30, 05:30, 07:30 LST respectively. The last one reveals the hourly rainfall amount during 07:30~08:30 local time in a 100 km by 100 km map of the area north of Seoul, which recorded a maximum of 100 mm per hour. The area in purple represents the area where rainfall amount exceeded 50 mm per hour. Source: KMA.

Storm forecasting for the next 0~6 hours, often called nowcasting, employs storm-tracking tools. The forecaster plays a vital role in capitalizing on mesoscale model products and the latest observations to derive warning decisions. It is necessary to maintain a total system approach that would take full advantage of both human and machine elements. Radar echo, as illustrated in chapter 1, tells us the direction and movement of severe weather systems nearby. The images are usually less than an hour old and are updated every few minutes. Each color in the image corresponds to certain rainfall rate threshholds, typically enhanced by red color for areas of heavy rain or heavy snowfall. The color scheme helps to identify rainfall patterns and signatures embedded in mesoscale convective systems. Radar imagery is often supplemented with satellite imagery and lightning strikes maps, which extend the radial views of approaching storms beyond the radar coverage that is typically bounded at 200~400 km.

Radar loop, or more specifically animated radar images in sequence, tells us where the rain band moves in and out, and how rapidly they develop or decay. The radar image is translated into rainfall rate with reference to and calibration against ground-based rain gauge observations. Using pattern recognition techniques with simple extrapolation of the shape of radar echo or equivalent precipitation pattern, one may quantitatively project the rain band movement and intensity for the next few hours. Due to the stochastic nature of storms and limitations in nowcasting tools, conceptual models are also used for the interpretation of mesoscale model outputs and latest observations. The conceptual models tell us that each storm cell tends to move along the vertical mean flow with a 1~5 km depth. Storms develop in large-scale environments where vertical instability and low-level moisture supply is supported with moderate shear and decay where such conditions cannot be sustained. It is difficult to trace the initiation of a new cell, even though the conceptual models shed some light on the favorable conditions for storm development. The prediction of storm intensity is an even more challenging subject.

Many centers develop their own tools for the automatic tracing of storm cells, and estimate the speed and direction of their movements. The System for Convective

Analysis and Nowcasting (SCAN) is a good example. It is an interactive analysis tool to monitor individual storm cells and to extrapolate their next positions up to a few hours ahead for a spot of interest that the analyst can select with a simple mouse click. Figure 9.3.3 illustrates an example from the JMA website. The spiral-shaped rainband accompanying typhoon Sinlaku on the lower panel is projected to evolve into the upper panel by nowcasting tools. The band concentrates toward the center of the typhoon while the whole system shifts eastward after 6 hours.

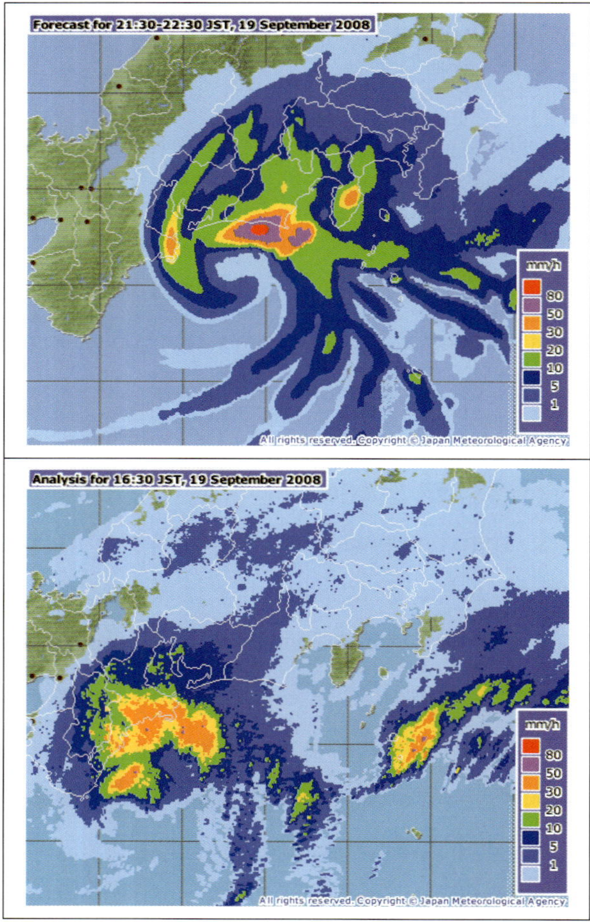

Fig. 9.3.3. Precipitation nowcast (upper) for 6 hours ahead using radar observation (lower) at 04:30

UTC on 19 September 2008 and sophisticated extrapolation techniques from the JMA website (http://www.jma.go.jp/en/radame/index.html?areaCode=204); the weakened typhoon Sinlaku traveled along the southern coast of Japan in an east-northeast direction.

To avoid unwelcome surprises from severe thunderstorms that arrive without warning, it is wise to first read the local forecast for the potential area of storm development. Then search the local website for the radar loop to confirm the nowcast.

10. Precipitation

10.1 Classification

Before making any adjustments to model precipitation, it is necessary to understand how models produce precipitation at a target region with the help of conceptual models. Model precipitation can be produced by frontal ascent, orographic lift, convection, tropical cyclones, and combination thereof.

Frontal ascent is frequently observed around an extratropical cyclone or a convergence zone with deformation flow. Figure 10.1.1 demonstrates an example of the frontal precipitation embedded in an extratropical cyclone. Along the western frank of the subtropical high, a stationary trough stretches from the southern China to the Yellow Sea. An upper-level short wave trough over northeastern China moving eastward supports the development of a low depression over the East Sea. Model precipitation is associated with the trailing cold front extending southwest from the low center. Models do a pretty good job at predicting the general features of frontal precipitation. Yet details such as intensity and areal coverage of the model precipitation need to be better adjusted to the observation. In the example, 3-hr accumulated observed precipitation amounts to 50 mm, which is 2.5 times of what the model predicted. In addition, the narrow band of pre-frontal precipitation over the middle of the Peninsula is not well captured in the model precipitation.

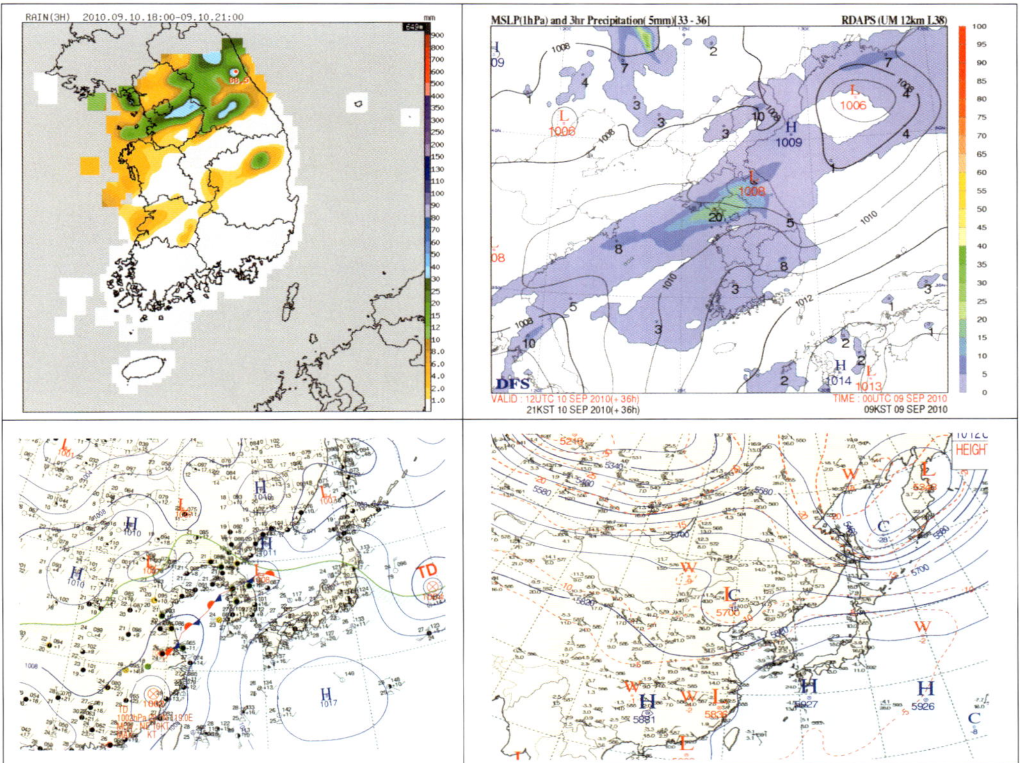

Fig. 10.1.1. Frontal precipitation in the UM model with a grid spacing of 12km and 38 levels along the trailing cold front stretching from the extratropical cyclone centered over the northern East Sea. The observed precipitation amount (upper left) is compared with the 33~36 hr forecast precipitation in units of mm overlaid with mean sea-level pressure (upper right), surface analysis with fronts, and 500 hPa analysis (lower panels) valid 21 LST on 10 September 2010. The front passed over the Korean Peninsula around 8 LST on 11 September 2010. Source: KMA.

Precipitation distribution associated with tropical cyclones, as shown in Fig. 10.1.2, is also fairly well simulated by models. The UM model predicted the spiral precipitation band of the typhoon Kompasu. It even simulates the precipitation band along the pre-convergence zone over the Yellow Sea at midnight on 31 August 2010. Generally, the accuracy of precipitation forecast along the tropical cyclone path critically depends on the model performance in predicting its position and intensity. In the examples above, the model fairly well simulates the evolution of the tropical cyclone, which in turn helps predict the associated precipitation.

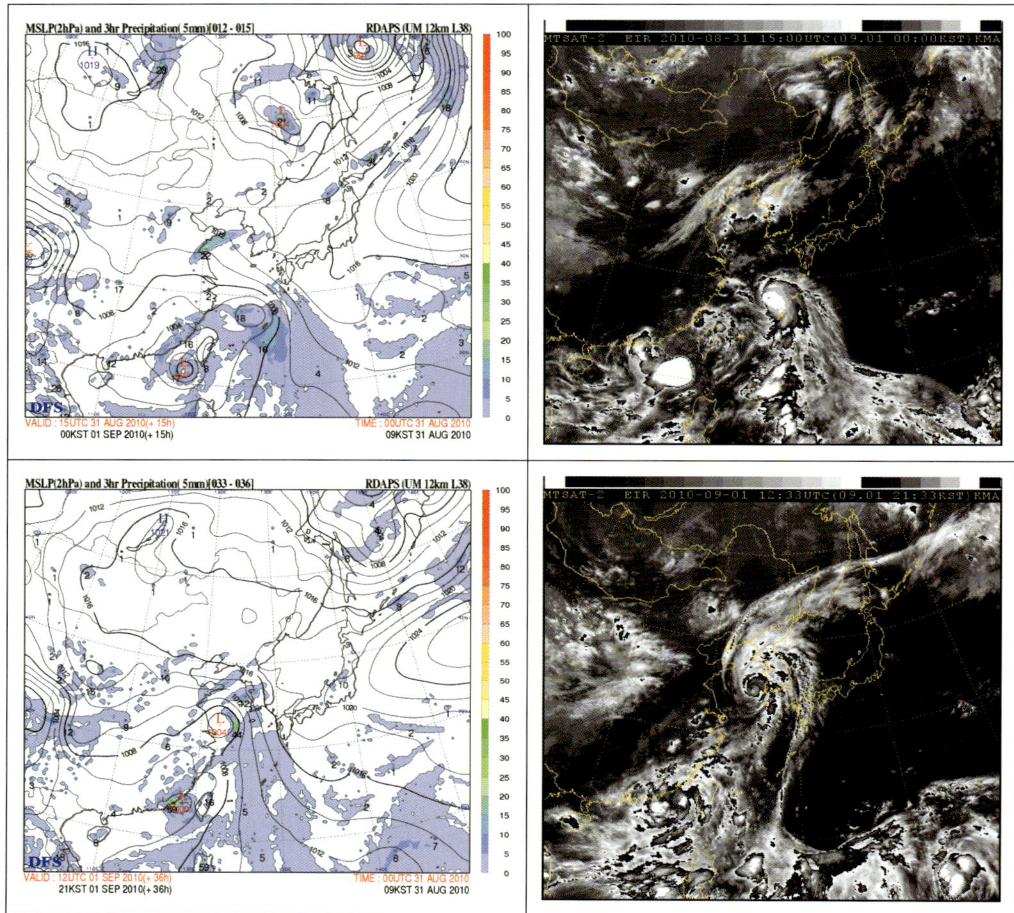

Fig. 10.1.2. Forecasted 3-hr accumulated precipitation in units of mm overlaid with MSLP (left) and enhanced infrared cloud image of MTSAT-2 at valid time. The UM Model with a grid spacing of 12km and 38 levels was used to derive the forecast from the initial state at 09 LST on 31 August 2010. The upper left panel is the 18-hr forecast simulating the precipitation band along the pre-convergence zone in front of the approaching typhoon Kompasu. The lower left panel is the 36-hr forecast for the precipitation influenced by Kompasu with its spiral bands. Source: KMA.

Other types of model precipitation patterns are not always easy to interpret without detailed investigation. Such cases are for the most part associated with the overrunning of warm air over cold surface in a weakly forced environment. An example along a stationary front is presented in Fig. 10.1.3. The area of ascent in the isentropic surface is identified by the wind vector pointing toward the low value of geopotential, indicating warm air stream overrunning the cold air beneath. In the figure, the model precipitation band between the southeast of Mongolia and northern Manchuria collocates to the region of the southwesterlies oriented toward the low pressure on a 280 K isentropic surface.

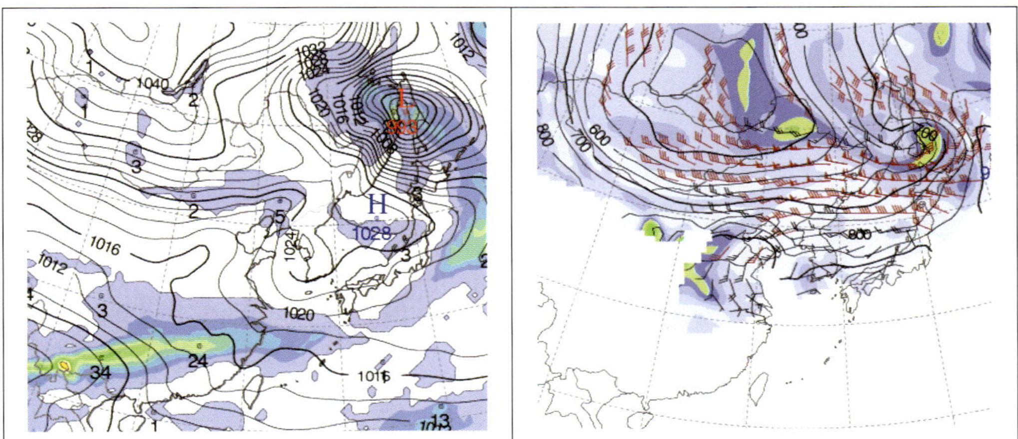

Fig. 10.1.3. An example of overrunning precipitation from the UM model with a grid spacing of 40km and 50 levels. (Left) D+4 forecast of 3-hr accumulated precipitation (color) and mean sea-level pressure in units of hPa (solid line), and (Right) D+4 forecast of isentropic pressure in units of hPa (solid line), wind barbs, and potential vorticity (color) in units of 10^{-6} m^2s^{-1} K at 280 K. Source: KMA.

Orographic enhancement of precipitation is indicated by low-level flow rising along the slope of mountain ridge. This type of precipitation is sensitive to the geography, wind direction and vertical stability. An example is given in Fig. 10.1.4. Sustained southerly wind flow brings a considerable amount of precipitation on the southern coast of the Korean Peninsula. Owing to the limited topographic resolution within the model, the model precipitation reaches only half of the observed near the central southern coast.

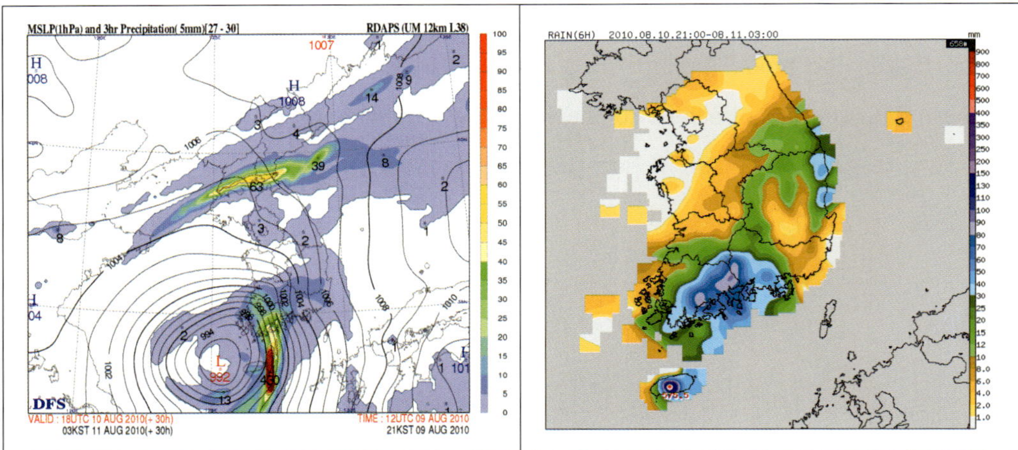

Fig. 10.1.4. Orographically enhanced precipitation interacting with an approaching tropical cyclone in the UM model with a grid spacing of 12km and 38 levels. (Left) The 27 to 30 hr forecast precipitation in units of mm is overlaid with MSLP in units of hPa: green color represents the precipitation rate of 40 mm/3hr. (Right) The model forecast is compared with the observed precipitation valid at 18 UTC on 10 August 2010: grey color represents the precipitation rate of 100mm/3hr. Source: KMA.

Model precipitation is produced by the microphysics (MP) and the convective parameterization (CP) schemes. The contribution of CP comes from the release of convective instability. It is noted, however, that part of MP precipitation is produced by the condensation of moisture redistributed by CP. A high-resolution non-hydrostatic model can even generate convective precipitation solely based on the grid-scale MP. Thus the division between non-convective and convective precipitations is somewhat arbitrary in a model. Figure 10.1.5 illustrates an example of their relative contributions. The model convective precipitation southeast of the low center grossly matches with the areas of observed rainfall and lightning, but is slightly displaced northward.

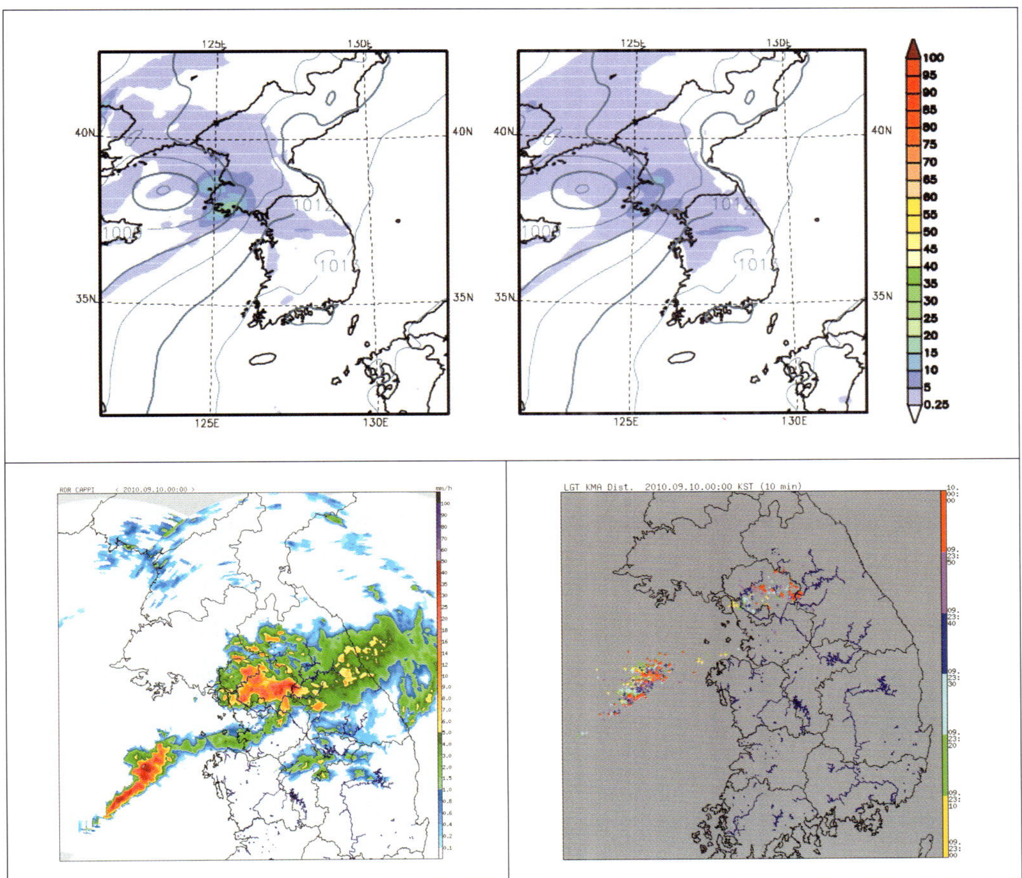

Fig. 10.1.5. 24 to 27 hr forecast of non-convective or microphysics precipitation (upper left) and convective precipitation (upper right) along a cold front in the UM Model with a 12-km grid spacing and 38 levels. They are compared with the composite radar reflectivity in units of mm/hr (lower left) valid at 00 LST on 10 September 2010. The red color represents the converted rainfall rate exceeding 20mm/hr. The lower right panel shows the distribution of lightning strikes with time of detection; red dot - 30 minutes earlier, pink dot - 60 minutes earlier, blue dot - 90 minutes earlier, light blue dot - 120 minutes earlier. Source: KMA.

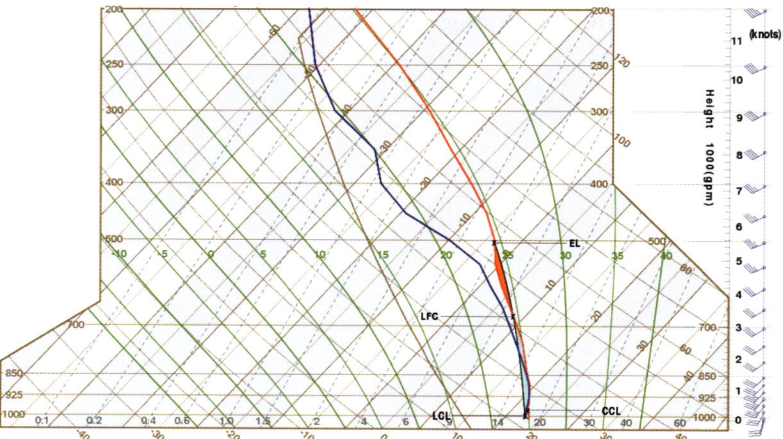

Fig. 10.1.6. Thermodynamic diagram taken near the southeastern corner of a low depression where both convective and non-convective precipitation occurs in the UM model forecast in Fig. 10.1.5. The profile is adjusted toward neutral instability through a convective parameterization process. Source: KMA.

Convective precipitation can be confirmed by the model sounding and instability indices. One should be mindful, however, that the model sounding after the convective adjustment often represents the state of saturation and may not directly correspond to the strength of potential instability at the time of active convection. The profile for convective precipitation in Fig. 10.1.6 is adjusted toward neutral instability in the layer between 500~1000 hPa through a convective parameterization process.

10.2 Role of convection

A major function of CP is vertical redistribution of moisture and heat to emulate release of convective instability by smaller-scale motions not represented by grid-scale motion. The CP suppresses the numerical instability that can be caused by anomalous convection.

CPs differ in convection triggering, vertical redistribution of heat and moisture by clouds, and closure condition for low-level mass flux. Characteristics of different CPs are summarized in Table 10.2.1.

Table 10.2.1. Comparison of convective parameterization schemes, reproduced from COMET (http://www.comet.ucar.edu/nwplessons/precipproclesson3/convparambackground.htm)

Scheme	Convection initiated by	Modifies by	Downdraft processes	Comment
Modified Kuo	Moisture convergence	Adjustment	No	No clouds
BMJ	Buoyancy/ moisture	Adjustment	No	No clouds
Kain-Fritsch	Subcloud layer convergence	Mass flux	Yes	Streamlined cloud model
Tiedtke	Moisture convergence/ PBL turbulence	Mass flux	Yes	Streamlined cloud model
Arakawa-Schubert	Destabilization rate	Mass flux	No	Models multiple cloud sizes
Grell	Destabilization rate/ CIN	Mass flux	Yes	Streamlined cloud model

The Kain Fritsch (KF) scheme often generates overly spotty patterns of convection, while the Grell and Arakawa Schubert schemes tend to produce aggregated forms of convections. The characteristics of model convection depend not only on the CP, but also on other physical processes interacting with moist physics. In general, mass flux-type convection schemes are more flexible than Kuo and Betts-Miller-Janjic (BMJ) schemes in meeting diverse synoptic forcings.

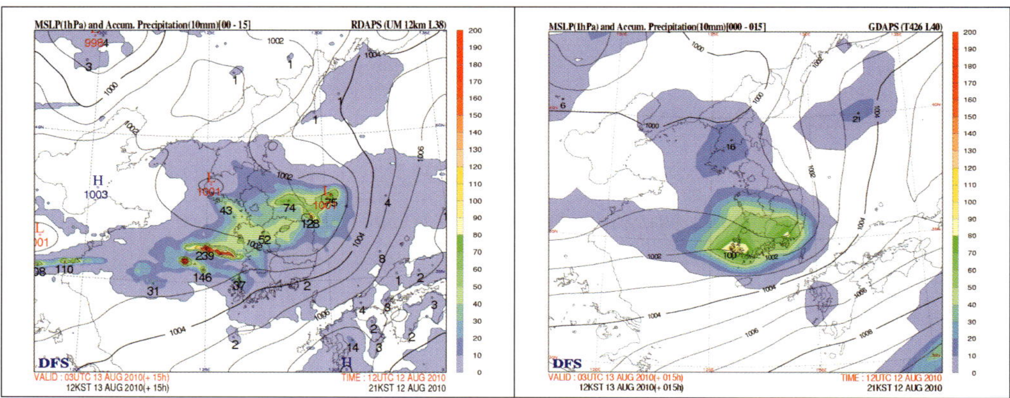

Fig. 10.2.1. 15-hr accumulated precipitations from models with the KF scheme (left) and Kuo scheme (right) in units of mm for a case of low-level jet with convergent flow. Source: KMA.

For the case of low-level convergent flow under conditionally unstable environments, both the KF scheme and the Kuo scheme produce organized convection along the line of confluence as illustrated in Fig. 10.2.1. On the other hand, they show considerably different behavior for the conditionally unstable environment with the weakly forced flow in Fig. 10.2.2. While the Kuo scheme does not generate convection, the KF scheme produces moderate convection. Understanding CP is essential to make necessary adjustments to the model convective precipitation.

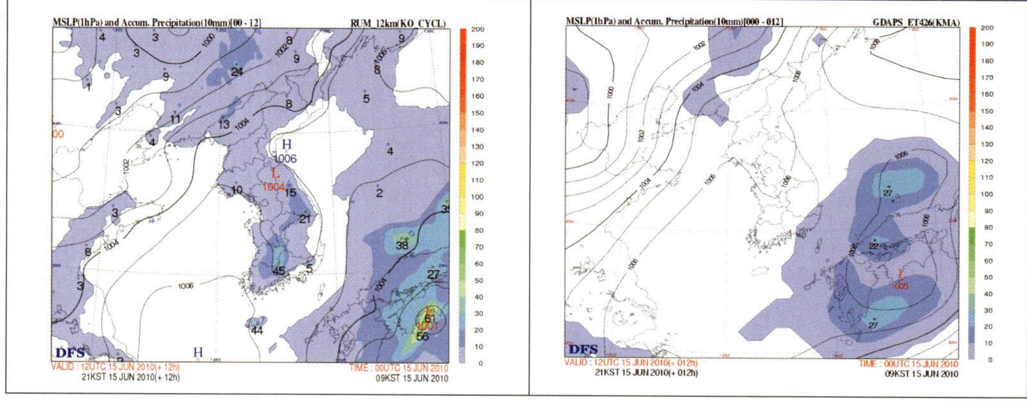

Fig. 10.2.2. 15-hr accumulated precipitations in units of mm for a case of mass convection with weak forcing, from models with the KF scheme (left) and Kuo scheme (right). Source: KMA.

The error behavior of model forecasts of convective precipitation is not always consistent. Let us consider the example in Fig. 10.2.3 which compares two computer model forecasts, WRF and Meso-scale Model version 5 (MM5), with ground observations. The two models differ in the coverage of rainfall from the beginning of the forecast period. The WRF model predicts the rainband position much further north than the MM5 model on D+1, which is closer to reality. Later on D+2, however, the MM5 model correctly simulates the southward migration of the band, while the WRF model does not. This example demonstrates that what starts out as a good forecast is not necessarily better than others at a later forecast time. It further implies the complexity of the problem involved in finding out which model forecast is better than others in a given situation.

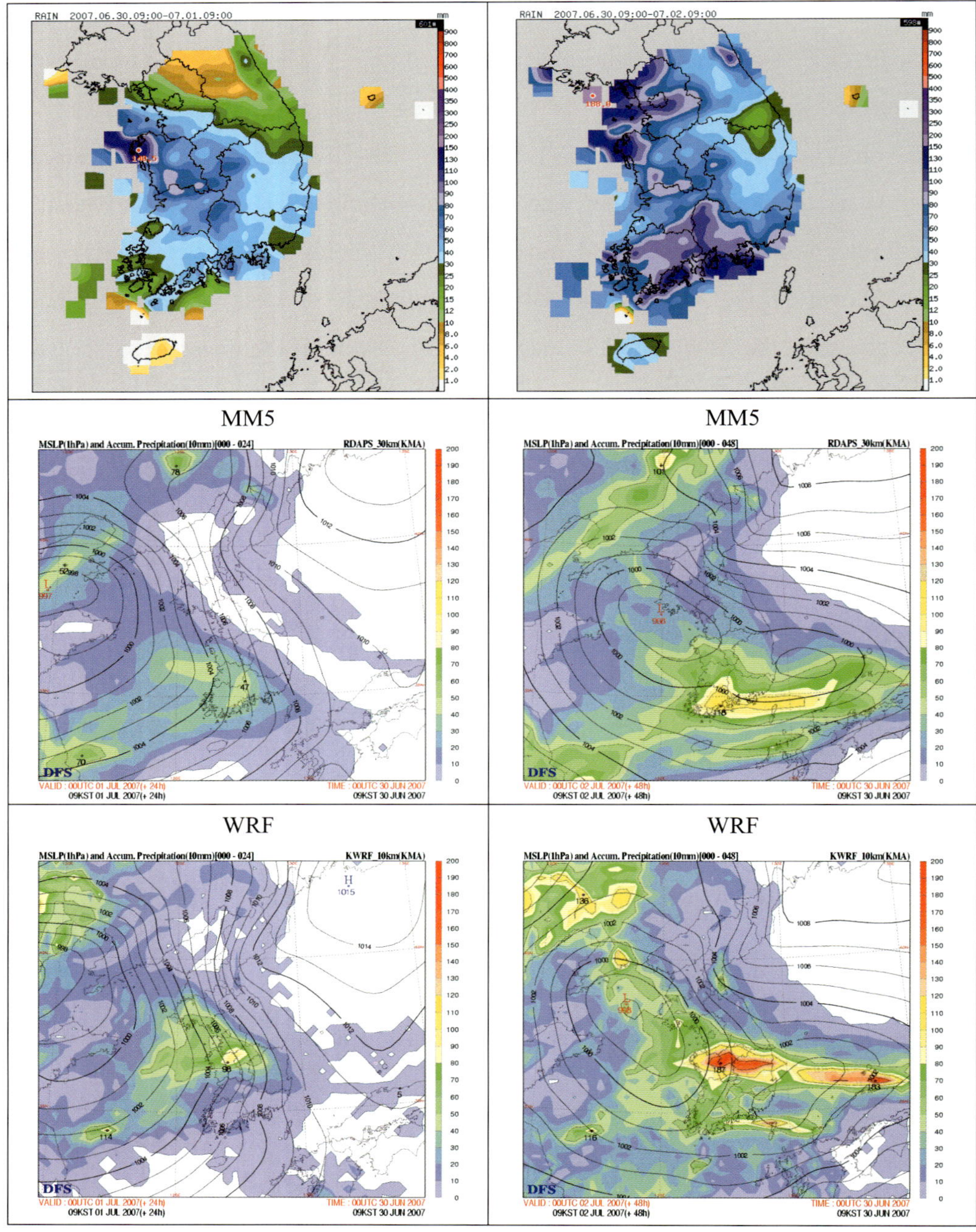

Fig. 10.2.3. Observed rainfall amount from surface rain gauges during 24 hours (09 LST 30 June to 09 LST 1 July 2007, top left), and during 48 hours (09 LST 30 June to 09 LST 2 July 2007, top right). The areas in grey are values exceeding 80 mm. The observed value is compared with the forecast rainfall from the two community numerical weather prediction models, MM5 (middle), and WRF (bottom) for the last 24 hours and 48 hours respectively. The two model forecasts

starting from the same set of initial observations show a large difference in terms of rainfall pattern and intensity. Source: KMA.

10.3 Correction of model precipitation

Precipitation intensity

Once the precipitation pattern is identified, one is ready to adjust the model precipitation in terms of spatial coverage, intensity, and timing. A common practice for correcting precipitation intensity is to trace the latest observation record of the associated weather system in the upstream region, and estimate the model bias in precipitation forecast. During summertime the rainfall rate over southern China provides a good hint on the inflation factor for the model rainfall over the Korean Peninsula. If such observation is not available, recent statistics of RMSE and mean error of model precipitation can be used instead for the precipitation range adjustment. The historical rainfall maximum provides an additional reference for extreme precipitation scenarios.

Model precipitation affected by orography requires particular treatment. Orographically enhanced precipitation strongly depends on the grid spacing in a model. High-resolution models produce more precipitation on the windward slope of a mountain ridge than low-resolution models if other conditions are equal.

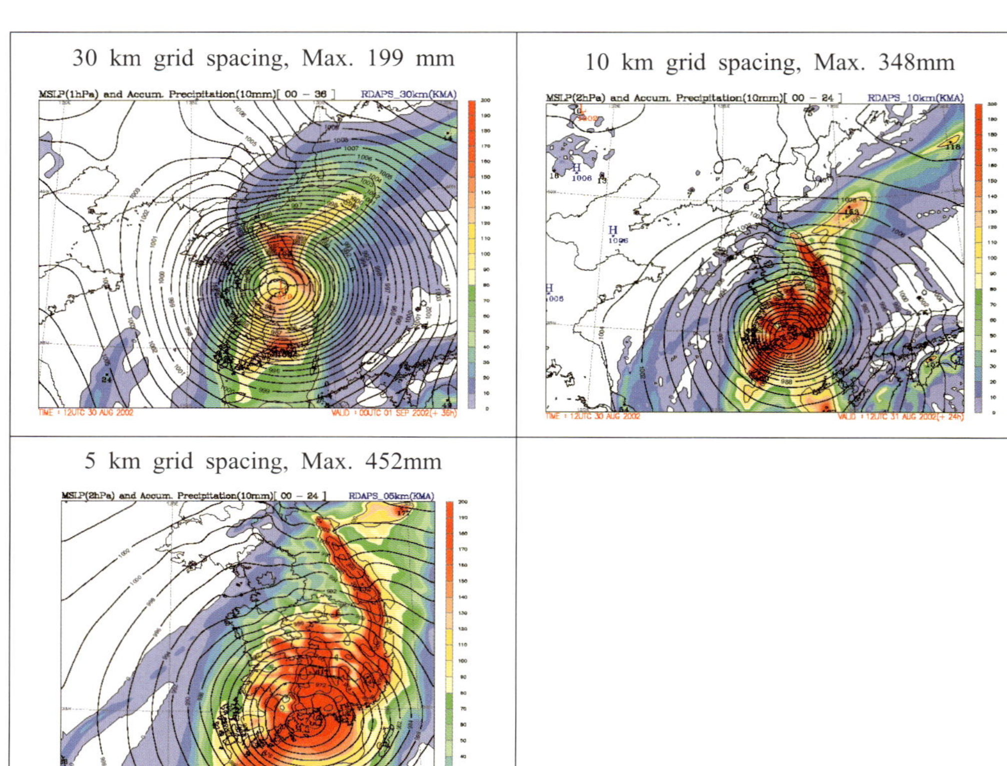

Fig. 10.3.1. Rainfall enhancement on the upslope of the central eastern mountains on the Korean Peninsula facing strong easterlies when Typhoon Rusa approaches the Korean Peninsula from the South Sea. 24-hr accumulated rainfall was produced from the MM5 meso-scale model with grid spacings of 30 km (upper left), 10 km (upper right), and 5 km (lower left), starting from the initial condition at 12 UTC on 30 August 2002. Source: KMA.

An example for the relationship between the model resolution and orographic enhancement is illustrated in Fig. 10.3.1 with the MM5 community meso-scale model. Near the central eastern mountain ridge of the Korean Peninsula facing the strong easterlies, the model with a 5-km grid spacing produced twice as much rainfall as the 30-km model.

Convective precipitation

Bias correction for convective precipitation is more challenging than for orographic precipitation, as the CP interacts with, and feedbacks to large-scale precipitation processes. Model convection can be classified into two types: underactive and overactive convection. Understanding each type provides tips for correcting model precipitation bias in the conditionally unstable atmosphere. Underactive convection in a model leaves a large amount of moisture at low levels. It is used to produce latent heating, which leads to rapid cyclogenesis. Induced secondary circulation with large-scale ascent motion supports stronger moisture convergence, and a positive feedback loop is established. On the other hand, overactive convection leaves little moisture behind. It tends to suppress precipitation from stratiform-type clouds and inhibit convection in the downstream region.

If other conditions are equal, model precipitation needs to be deflated for underactive convection and inflated for overactive convection. Some convective parameterization schemes generate anomalously large amounts of convective precipitation in a very localized area, as demonstrated in Fig. 10.3.2. The model thermodynamic diagram indicates lower-level convection with cloud top below 600 hPa. However, the model produces an anomalously large amount of precipitation, i.e., 222 mm per 3 hours, while the observation only records 45 mm. The model failure is called a point storm, and no counterpart is found in nature. It occurs mostly in the summer hemisphere under a humid and conditionally unstable environment.

Some non-hydrostatic meso-scale models do not include CP, and MP instead generates grid-scale convection. It takes time for the precipitable clouds to spin up, which results in delay of precipitation during the early hours of model forecast.

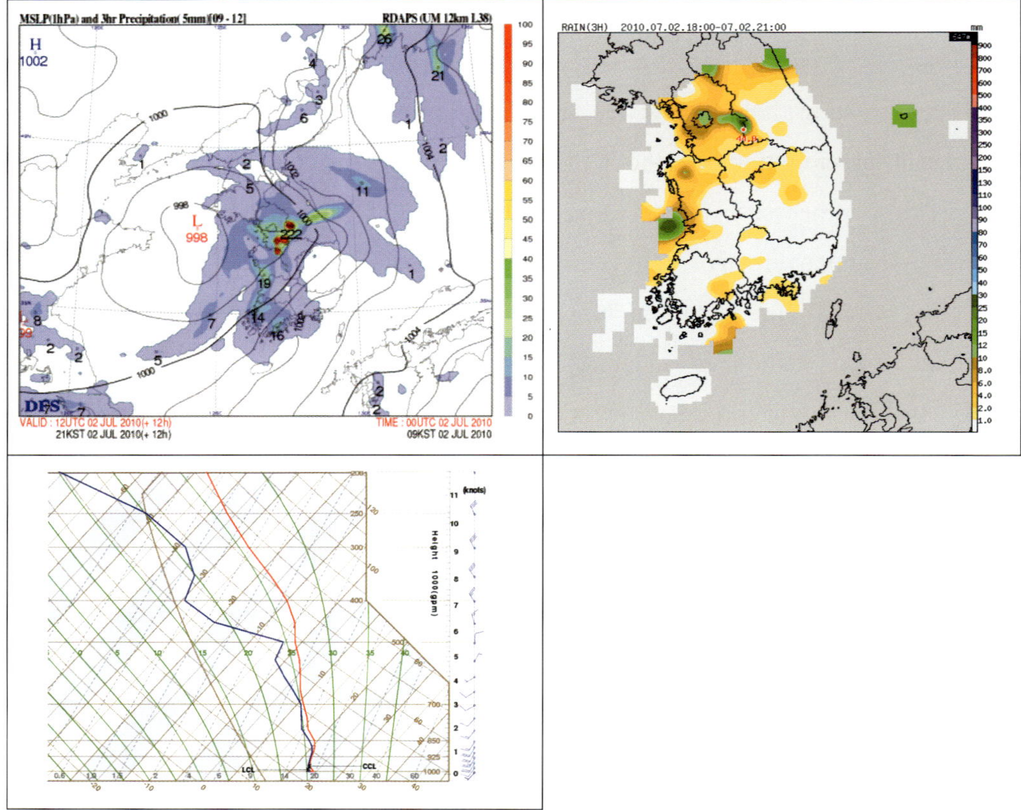

Fig. 10.3.2. Example of a point storm from the UM model with a 12-km grid spacing and 38 levels. (Upper left) 12-hr forecast of 3-hr accumulated precipitation (color) in units of mm and mean sea-level pressure (solid line) in units of hPa, (Upper right) observed 3-hr rainfall, and (Lower left) thermodynamic diagram valid 21 LST on 2 July 2010. Source: KMA.

Precipitation area

The spatial coverage of precipitation in a model is somewhat arbitrary due to the cut-off values imposed on contours of minimum precipitation. An example from the UM model with the NCAR graphics is provided in Fig. 10.3.3. The precipitation coverage varies with the different cut-off values. The observed coverage better resembles the forecasted coverage with the cut-off value of 1 mm. Many centers use different cut-off values that may even vary with season and precipitation period. It is left to the users to familiarize themselves with the cut-off value and associated precipitation map. Users should pay more attention to the gross features in the large-scale precipitation pattern with moderate or higher intensity than on the exact boundary of light precipitation.

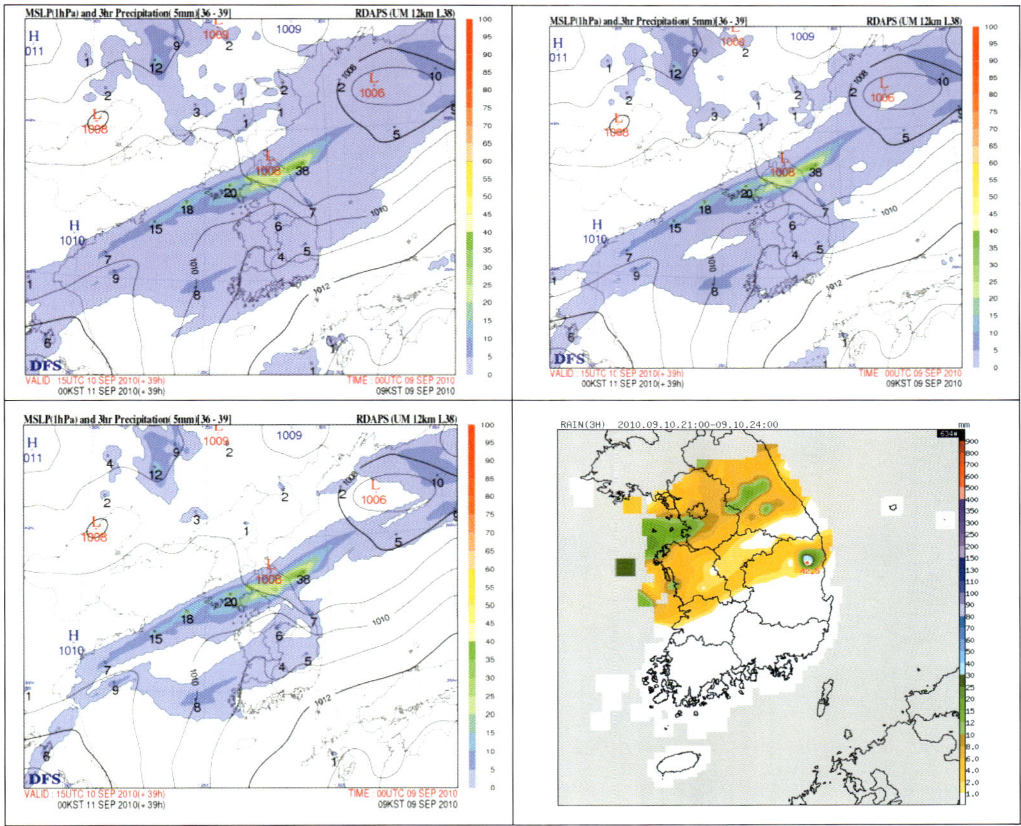

Fig. 10.3.3. Forecast precipitation during 36 and 39 hour with varying cut-off values: 0.1mm (upper left), 0.3 mm (upper right), and 1.0mm (lower left). Observed precipitation is shown in the lower right panel valid 24 LST on 10 September 2010. The UM Model with a 12-km grid spacing and 38 levels was used for the precipitation forecast starting from the initial condition at 09 LST on 9 September 2010. Source: KMA.

10.4 Ice-phase precipitation

MP in a model can produce ice-phase precipitation such as snow, ice pellets, graupel, and sleet through a cold-rain process called the Bergeron - Findeisen process, collision and coalescence. Model snow is less predictable than model rain as additional complexity is involved in ice-phase physics. Most advanced models predict the amount of ice-phase precipitation with varying degrees of complexity. An example is illustrated in the upper panel of Fig. 10.4.1. The 48-hr model forecast reveals a precipitation band along the cold front embedded in the low centered over Manchuria.

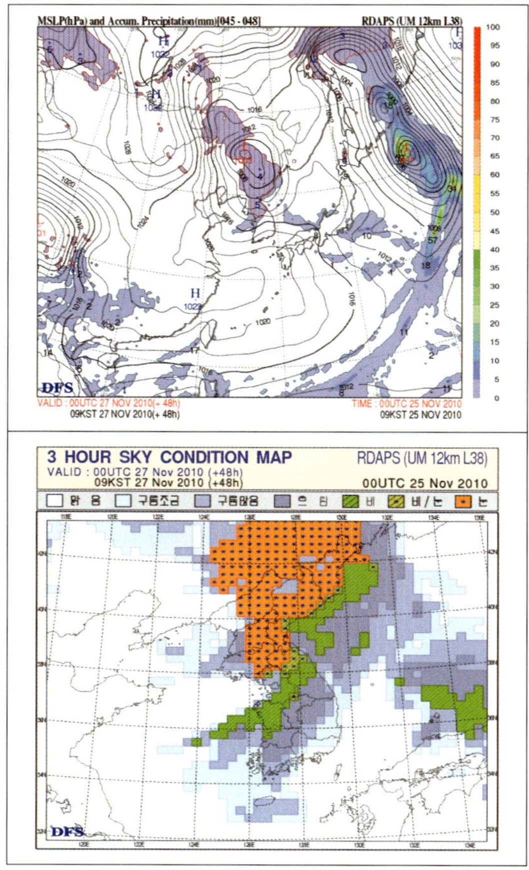

Fig. 10.4.1. Forecast of liquid and ice-phase precipitation during 45 and 48 hours in the UM model

with a 12-km grid spacing and 38 levels, starting from an initial state at 09 LST on 25 November 2010. (Upper) 3-hr accumulated precipitation and snow in units of mm are filled with colors and slashes respectively. (Lower) various types of precipitation, as a function of temperature and relative humidity at surface (Matsuo and Sasyo, 1980), are filled with colors: snow in orange, rain in green, and mixed precipitation in light green. Source: KMA.

Precipitation type

Model precipitation near freezing temperature can be further classified into various precipitation types with the help of conceptual models. The lower panel of Fig. 10.4.1 shows an example of precipitation type forecasts, determined by a simple formula proposed by Matsuo and Sasyo (1980). The formula is derived from the observed types of precipitation scattered on a 2-dimensional plane consisting of surface temperature and surface relative humidity, as illustrated in Fig. 10.4.2. Evaporative cooling in the unsaturated environment competes with sensible heating by conduction, as snowflakes fall through the warm layer beneath the clouds. Thus the chance of snow increases as relative humidity decreases. Snow is possible at 4 and 6 degrees above freezing if relative humidity is below 70% and 50% respectively. Below 2 degrees above freezing, most precipitation is in the form of either snow or sleet. Mixed precipitation is possible at surface temperatures between 2 to 4 degrees. In this temperature range, the precipitation type again depends on the relative humidity: if relative humidity increases above 85%, snow is unlikely, and chance of sleet decreases as relative humidity increases. Based on the conceptual model, snow in the lower panel of Fig. 10.4.1 only appears to the north of 38°N. The snow boundary is grossly comparable to that directly derived from the model microphysics except for some disagreement near the central part of the Korean Peninsula.

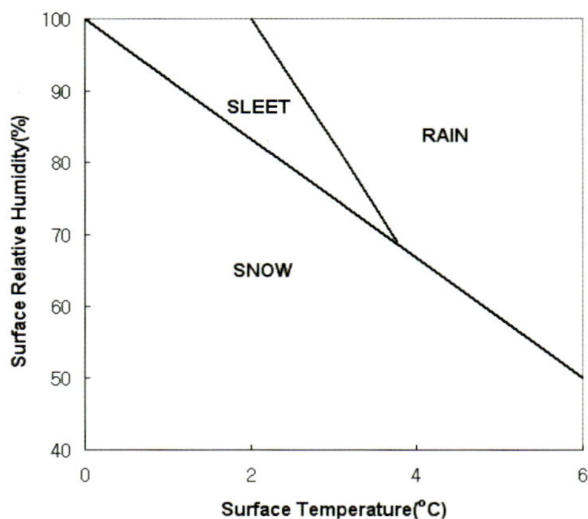

Fig. 10.4.2. Precipitation type on a plane of surface temperature and surface relative humidity (Matsuo and Sasyo, 1980).

The conceptual model in Fig. 10.4.2 has obvious limitations as it does not take into account the complex vertical structure in the cloud medium. The forecast sounding in the target area provides an additional clue for the possible precipitation type in the given synoptic environment. The precipitation type is sensitive to the freezing level (or zero Celsius) and wet-bulb zero temperature line, as demonstrated in Fig. 10.4.3. Three model thermodynamic diagrams are shown for the three selected points that represent forecast areas of snow, mixed, and rain respectively for the case in Fig. 10.4.1. While the surface temperatures are near freezing for all three soundings, the wet-bulb temperature for the moist layer (950~700 hPa) shows marked difference among the soundings. The wet-bulb temperature of the layer is only 2°C for snow at Pyongyang, but increases to 5~7°C for mixed precipitation and rain at Seoul and Daejeon respectively.

If a vertical temperature profile crosses the zero isotherm one time or more, various precipitation types can occur (Bourgouin, 2000). When the profile is well below the zero isotherm, snow is the only possible form of precipitation. When precipitation falls through the warm layer above the subfreezing layer near the surface, both freezing rain or ice pellets are possible. When most of the profile is above the zero isotherm, most of the precipitation is in the form of rain. The predictability of precipitation type is limited by the poor vertical resolution and sparse observation of moisture in the model boundary layer. Comparison of the latest model soundings with the latest observations is essential to adjust the model precipitation type.

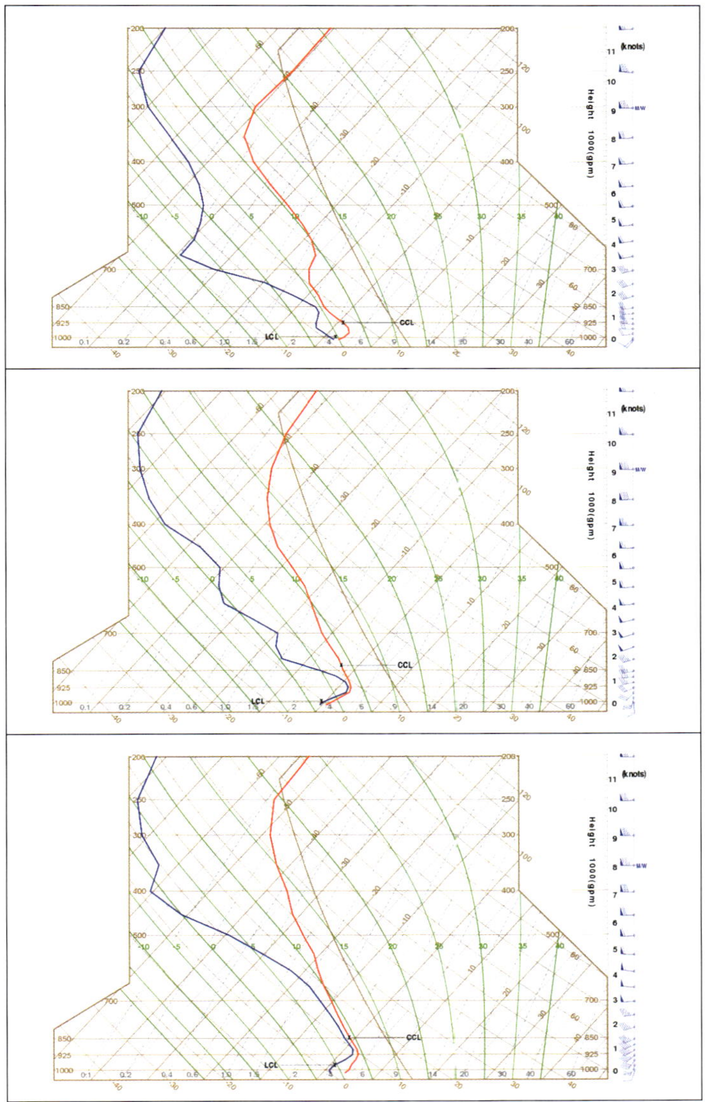

Fig. 10.4.3. Model thermodynamic diagram for snow at Pyongyang (upper), near-mixed precipitation at Seoul (middle), and rain at Daejeon (lower) corresponding to the forecast case in Fig. 10.4.1. Seoul is located about 200 km north of Daejeon, and about 230 km south of Pyongyang. Source: KMA.

Snow depth

Model snowfall amount can be converted into snow depth by considering snow wetness and ground temperature. Snow wetness is typically measured by the ratio between the snow depth and its equivalent in liquid form. In general, moisture content in the air increases with temperature, and the snow and rain ratio decreases with latitude. In synoptic systems, snow associated with a warm conveyor belt has higher snow and rain ratio than snow associated with a cold one. According to the snowfall report for the United States in Fig. 10.4.4, the typical snow and rain ratio is around 10:1, but varies from 2:1 to 45:1 with different synoptic conditions.

The ground temperature determines the phase of the snow on the ground. The heat of fusion during the phase transition makes additional contribution to the surface energy budget. The ground temperature decreases down to freezing level when snow falling through the warm layer melts down, and increases toward zero Celsius when rain drops freeze on the cold surface.

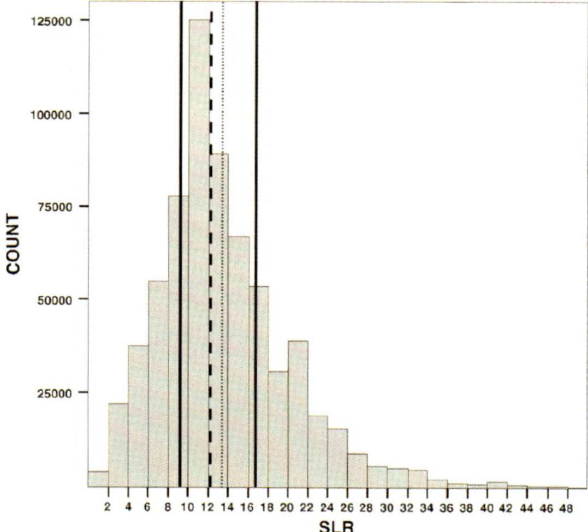

Fig. 10.4.4. Histogram for snow-to-liquid-equivalent ratio found in the United States. Solid lines represent 25th and 75th percentiles, the long-dashed line represents the median, and the short-dashed line represents the mean (Baxter et al., 2005).

Lake-effect snow

When the arctic cold air flows over warm water with strong northwesterlies, heavy snow bands often develop along the direction of the prevailing low-level wind. The vertical temperature gradient near the surface layer supports the convective transport of heat and moisture from the warm water up to the capping inversion near the top of the cold air mass. The strength and width of the convective cloud bands depend on the speed and depth of the unidirectional low-level wind, and on the temperature contrast between the air and water beneath.

Forced by high topography or by frictional convergence, precipitation processes are activated in the shallow convective clouds near the shore. At times several inches of snowfall in a day are recorded downstream of coastal zones. This often persists for several days until the ambient wind shifts its direction. Light snow flakes reach a few hundred kilometers downstream far inland. This phenomenon is called "lake-effect snow", and is illustrated in Fig. 10.4.5. There are several locations where cold air frequently impinges into the relatively warm water body during cold air outbreaks. Northeastern regions of major continents are favorable for lake-effect snow, including the southeastern edge of the Great Lakes, the eastern islands of Canada, the west coast of Korea, and the west coast of Japan.

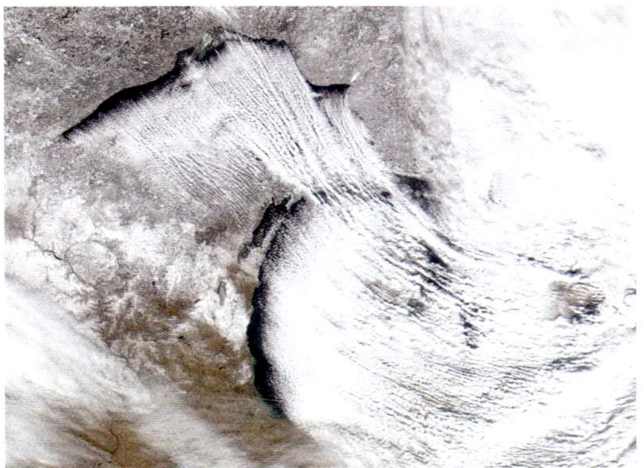

Fig. 10.4.5. Examples of lake-effect snow that occurs downstream of warm bodies of water with arctic

cold winds blowing from the polar regions in winter. Fine-scale snow bands are observed southeast of lakes Superior and Michigan, United States, taken on 5 December 2000 from SeaWiFS NASA. The image is available from http://apod.nasa.gov/apod/image/0412/lakeeffect_seawifs_big.jpg. Source: NASA.

Two case events of lake-effect snow on the western coast of the Korean Peninsula are demonstrated in Fig. 10.4.6. Both events are associated with cold arctic air, intruding relatively warm water bodies, with the southeastward expansion of the Siberian High. Streets of shallow convective clouds develop in line with the lower-level ambient wind. The model predicts fairly well the convective precipitation over the Yellow Sea, and the lake-effect snowfalls on the western coast of the Korean Peninsula. Convective clouds often interact with the upper-level trough, and enhanced precipitation is brought downstream further with the middle- and upper-level flow. The model was able to simulate the extended area of snowfall inland for the case of the approaching upper-level trough in the lower panels of the figure.

The necessary conditions for lake-effect snow include strong and unidirectional wind from the polar region, inversion, and saturation with air-sea interaction. The inland sounding near the Yellow Sea indicates that the lower layer (750~1000 hPa) is saturated along with strong northwesterlies and inversion at 750 hPa, preserving the properties of the convective boundary layer below. The model sounding also captures these ingredients favorable for lake-effect snow, as demonstrated in Fig. 10.4.7.

Even though the model was able to predict the general areas of shallow convective clouds and drifting snow downstream, the timing, intensity, and coverage of snow often deviate from the observed. The organization of cloud streets is crudely represented even in a high-resolution model.

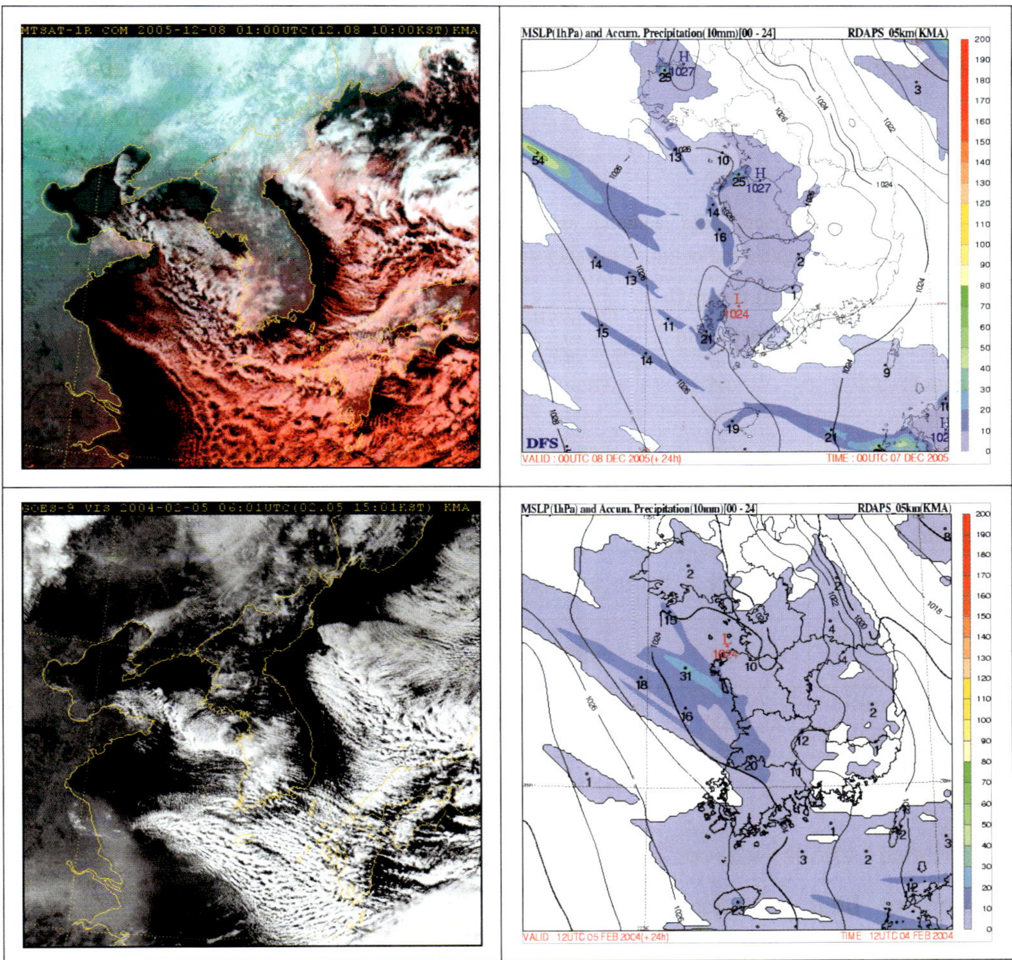

Fig. 10.4.6. Examples of lake-effect snow with a strong cold surge (upper panels), and a weak cold surge but supported by an approaching upper-level trough (lower panels). The upper panels are visible imagery from MTSAT-1R and the corresponding 24-hr accumulated precipitation in units of mm valid 10 LST on 8 December 2005, derived from the MM5 model with a 5-km grid spacing and 40 levels. The lower panels are visible images from GOES-9 and the corresponding 24-hr accumulated precipitation valid 15LST on 5 February 2004. The vortex shape organization of convective clouds over the Yellow Sea in the lower left panel indicates the interaction of convective clouds with the approaching upper-level trough, which transports lake-effect snow further inland with the middle-level flow. Source: KMA.

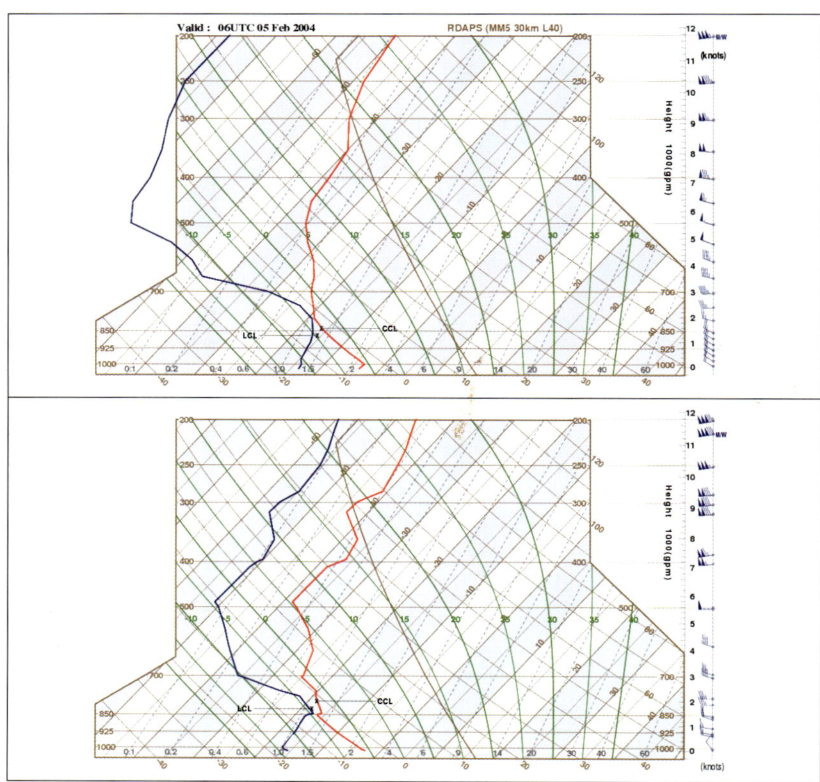

Fig. 10.4.7. D+1 model thermodynamic diagram near Seoul, 50 km inland from the Yellow Sea, derived from the MM5 model with a 5-km grid spacing and 40 levels (upper), and verifying analysis (lower), for the case of the weak cold surge in the lower panels of Fig. 10.4.6. Source: KMA.

11. Clouds, Fog, and Dust

11.1 Cumulus and stratus

Clouds form under a condition of large-scale lifting or convergence of moist air. They dissipate as a result of dry air mixing or compressive warming. In order to predict cloud water and ice in the atmosphere some computer models use prognostic equations, while others use simplified diagnostic relations. Model clouds evolve under advective processes and various sources and sinks including: condensation, evaporation, sublimation, conversion into precipitation and fallout.

Clouds are highly local and sensitive to atmospheric conditions including moisture field, vertical instability, and boundary layer characteristics. While the spatial and temporal resolution of contemporary models is not sufficient to represent location, size, and depth of individual clouds, synoptic-scale parameters in the model provide supplementary information for cloud forecast. Model dew-point depression (DD) is a measure of saturation of moist air, and indicates the coverage of model clouds. DD less than 2 degrees indicates overcast conditions, and DD between 2 and 4 degrees imply broken clouds at low levels around 850~700 hPa. Figure 11.1.1 presents the general relationship among cloud fraction, temperature, and dew point depression (Choi et al., 2005).

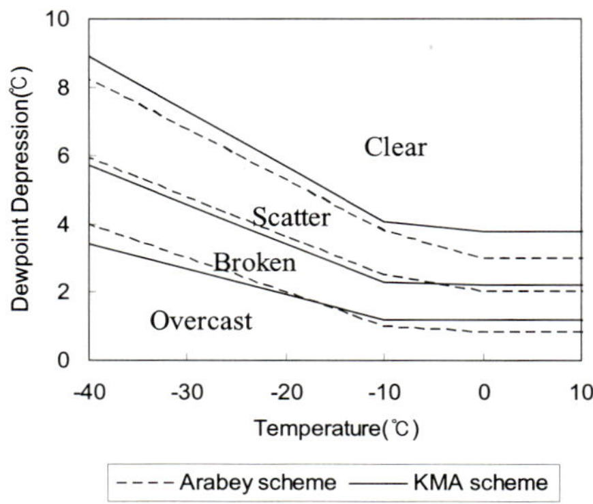

Fig. 11.1.1. Arabey scheme and KMA scheme to determine cloud coverage from temperature and dew point depression. “Overcast” refers to cloud fractions over 7.5; “broken”, between 5 and 7.5; “scatter”, between 2.5 and 5; and “clear”, below 2.5. The solid line and dashed line are for the KMA scheme and the Arabey scheme, respectively (Choi et al., 2005; Chernykh and Eskridge, 1996).

The cloudy region grossly matches with the areas of ascending motion. It is dynamically collocated with the downstream region of positive potential vorticity (PV) anomaly in the upper troposphere. The cloud bands over Mongolia in Fig. 11.1.2 correspond to the positive IPV anomaly at the 310 K isentropic surface at 9 LST on 17 November, and moved southeastward, reaching the Korean Peninsula after 48 hours.

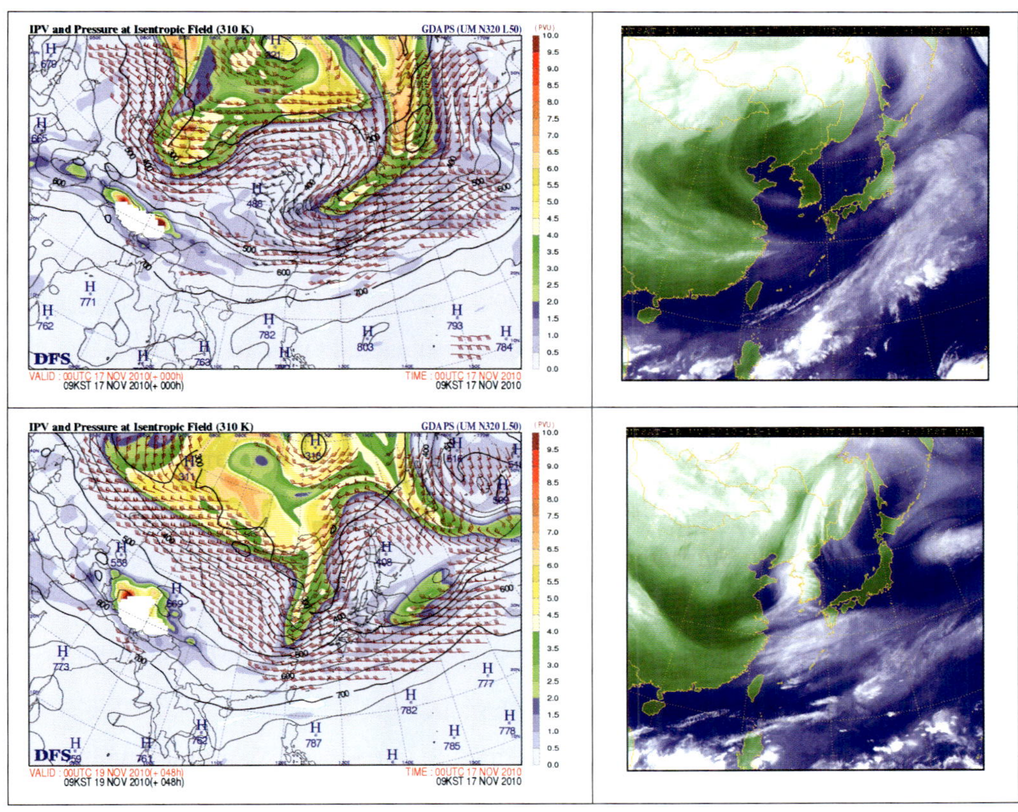

Fig. 11.1.2. Analysis (upper left) and D+2 forecast (lower left) of isentropic potential vorticity (IPV) in units of 10^{-6} m^2s^{-1} K (color), pressure in units of hPa, and wind at the 310 K isentropic surface from the UM model with a grid spacing of 40km and 50 levels, and verifying MTSAT-1R water vapor images valid 09 LST on 17 November, and 09 LST on 19 November 2010 (right panels). Green color represents the IPV value between 3.5~4×10^{-6} m^2s^{-1} K. Source: KMA.

Convective precipitation in a model is produced from the convective parameterization scheme, and is used to discriminate cumuliform clouds from other types of clouds. Model convection is initiated with cloud depth exceeding a certain threshold. For instance, cloud depths of 4 km over land and 1.5 km over the sea are needed to initiate convective precipitation in the UM model, except for a cold cloud top below -10°C where precipitation is allowed with cloud depth exceeding 1 km.

11.2 Fog and visibility

It is difficult to discern ground fog from low-level stratus in a model without reference to supplementary observations and conceptual models. Stratiform clouds are favorably found in the area of large-scale ascent with an approaching extratropical cyclone. An anticyclonic flow region in the vicinity of the same cyclone often leads to advection-radiation fog. It would be safe to check first the synoptic conditions conducive for low-level clouds, and then further explore the possibility of ground fog.

Contemporary models poorly represent the patch of advection-radiation fog, as the models cannot resolve the local details of surface characteristics and moisture source from river streams and ponds. On the other hand, they can fairly well represent coastal fog organized by large-scale advection of warm moisture from the open ocean, as demonstrated in Fig. 11.2.1. The temperature in the middle of the Yellow Sea is around 15°C, which is about 5 degrees warmer than the west coast. Low-level flow convergence over the western coast is evident where the warm moisture from the Yellow Sea condenses into fog. This is confirmed by visibility values less than 0.5 km along the west coast.

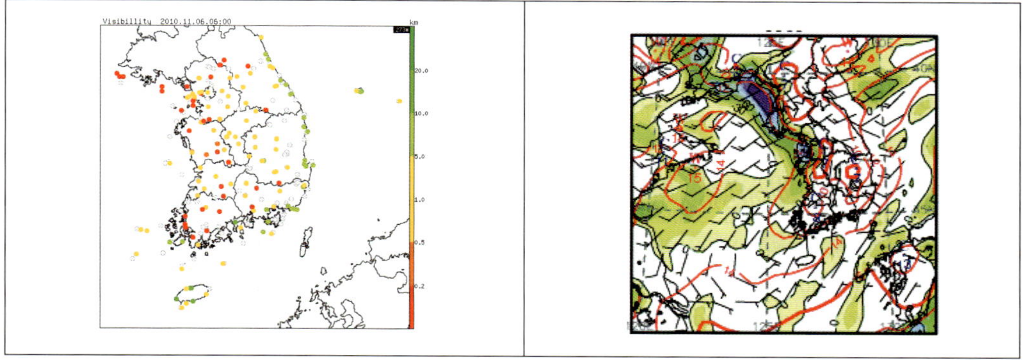

Fig. 11.2.1. Observed visibility in units of km valid 06 LST on 6 November 2010 (left), and D+1 forecast of wind feather, flow convergence (color filled), and wet bulb temperature (red) in units

of kt, $10^{-6}s^{-1}$, and Celsius, respectively at 1000hPa valid 09 LST on 6 November 2010 (right). The red, yellow, and green dots indicate visibility less than 0.5 km, between 0.5 and 5.0 km, and over 10 km respectively. The green color filled area denotes the convergence between $30\times10^{-6}s^{-1}$ and $60\times10^{-6}s^{-1}$. The UM model with a grid spacing of 12 km and 38 levels is used for the forecast. Source: KMA.

The model thermodynamic diagram in Fig. 11.2.2 indicates the saturation of moist air leading to advection-radiation fog with low-level westerlies from the Yellow Sea of Korea.

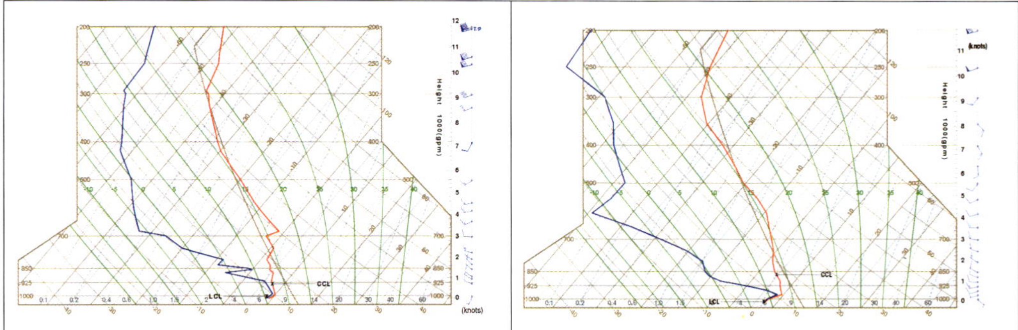

Fig. 11.2.2. Observed (left) and forecasted (right) thermodynamic diagram around Seoul, valid 09 LST on 06 November 2010 in correspondence with Fig. 11.2.1. Source: KMA.

Visibility is more a challenging element to forecast than fog or stratus coverage, as it depends not only on the existence of condensed hydrometeors but also on their size distribution. Even though the size distribution of various hydrometeors is prescribed in the microphysics of modern models, their prediction is an area of active research. For the time being, statistical guidance and conceptual models are as useful as the dynamic models for forecasting visibility.

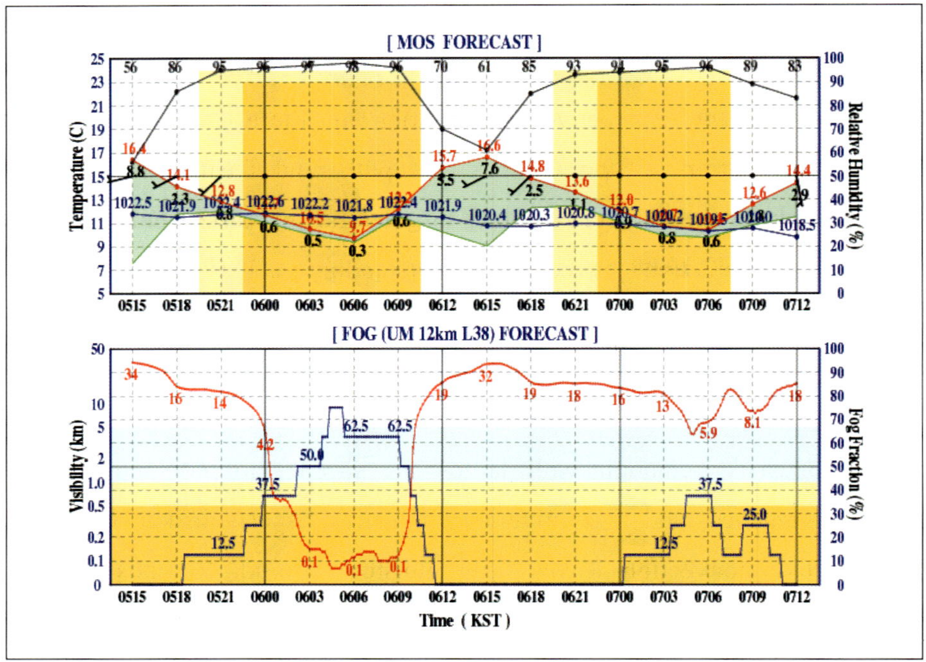

Fig. 11.2.3. (Upper) Statistical guidance of relative humidity in units of percent (black), temperature (red), and dew-point temperature (green) in units of Celsius at 2 m above ground; green area indicates dew-point depression, and yellow area represents for relative humidity exceeding 95%. (Lower) Model forecast for visibility in units of km (red) and fog fraction in units of percent (blue). The UM model runs from the same initial state as used in Fig. 11.2.1. Source: KMA.

An example of forecast guidance for visibility is presented in Fig. 11.2.3, based on the UM Model. The upper panel shows the forecasted relative humidity at 2m above ground after the model bias has been corrected using the MOS technique. The lower panel illustrates the forecasted visibility and fog fraction derived from the forecasted relative humidity and distribution of hydrometeors from the UM model.

11.3 Transboundary transport of tracers

Most desert areas are under the influence of a subtropical high-pressure belt that covers central Asia, the Persian Gulf, North Africa, inland Australia, and the northwest of South Africa. The subsiding air motion field warms the air beneath, and contributes to the maintenance of a stable boundary layer near the earth surface. High-pressure systems can also be supported dynamically downstream of major mountain ranges such as the Tibetan Plateau, the Rockies, and the Andes, where again, desert areas are found. Rain is hard to get under conditions of subsidence, inversion, and lack of a moisture source.

Floating dust in large-scale areas covering part of a country cause transboundary problems downstream of a source region. For instance, the Hwangsa or Asian yellow dust, a kind of blowing dust originating from the Asian deserts, travels with the downwind flow to eastern China, Korea, Japan, and sometimes even to North America across the Pacific. Once lifted with the gust wind ahead of a cold front, the Asian yellow dust moves with mid-level winds (about 700 hPa) and subsides along the poleward flank of a migratory high-pressure system. The trajectory of dust particles depends not only on the translation speed of weather systems but also the wind circulation embedded within. It is not easy to derive their trajectories from weather maps, and consultation with a weather expert is highly recommended.

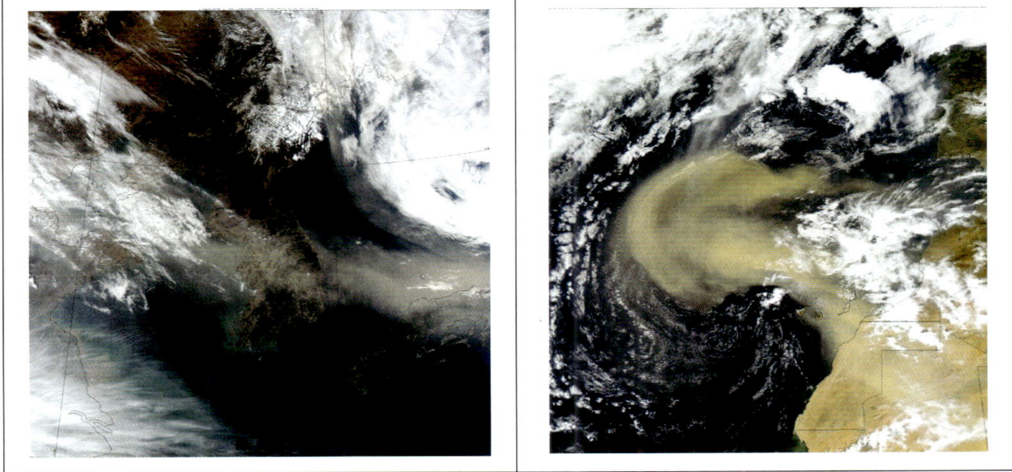

Fig. 11.3.1. (Left) Hwangsa or Asian yellow dust, as captured by MODIS at 04:35 UTC on 8 April

2006, frequently occurs in spring when dust from China and Mongolia are transported downstream by strong northwesterlies. Source: KMA. (Right) A sandstorm blowing off the northwest African desert expands to hundreds of thousands of square miles of the eastern Atlantic Ocean with Saharan sand. The image is taken from http://antwrp.gsfc.nasa.gov/apod. Source: NASA.

Model simulations are widely used to trace and predict the trajectory of pollutants of either natural or industrial origin, and their concentration. Hereafter the interpretation of dust concentration forecast is taken as an example for the transboundary transport of Asian yellow dust. A similar approach can be applied to other pollution problems including chemical spills, volcanic ash, and radioactive pollutants.

Three sources of uncertainty are involved in the prediction of dust concentration: the generation rate in the source region, the en route wind forecast, and the physical process of pollutants interacting with atmospheric particles. The source region is covered mostly by desert or rugged terrain where little ground observation is available. Satellite remote sensing demonstrates limited ability to discriminate between dust and clouds, and thus satellite radiance data assimilation is used for analyzing dust concentration.

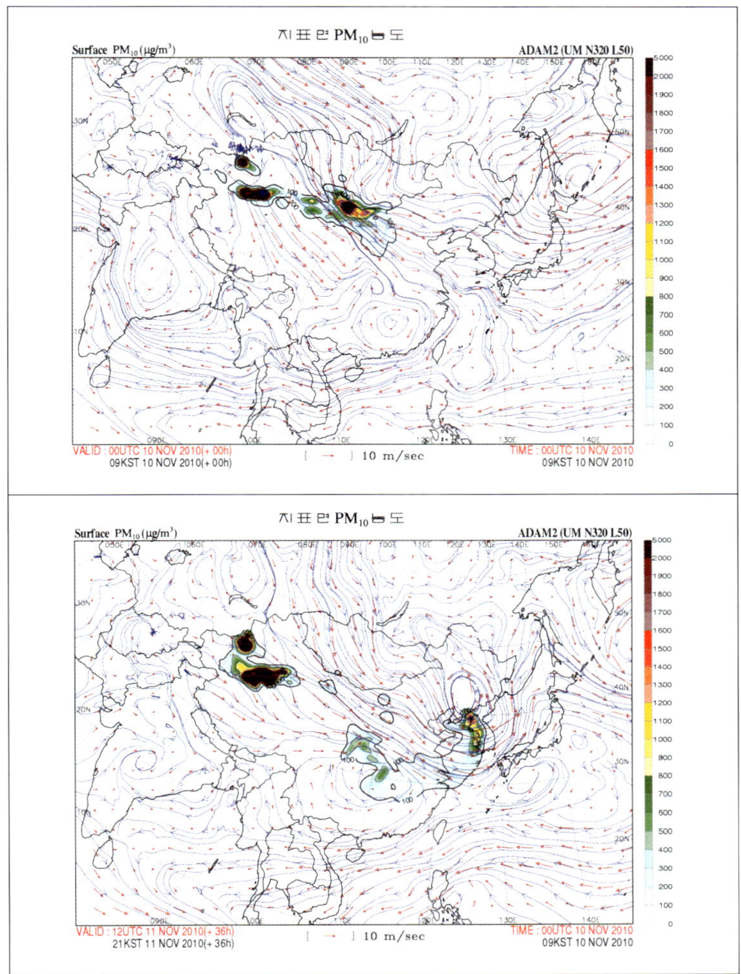

Fig. 11.3.2. Initial analysis at 09 LST on 10 November 2010 (upper), and 36-hr forecast for surface dust concentration in units of $\mu g/m^3$, the streamline and wind vector at 850 hPa (lower), derived from the Asian Dust Aerosol Model version 2, ADAM2 (Park et al., 2010). Source: KMA.

Most operational models start with the proxy dust concentration that is derived from the boundary layer processes responsible for lifting the dust into the atmosphere. Figure 11.3.2 presents the forecast dust concentration from the Asian Dust Aerosol Model version 2, ADAM2 (Park et al., 2010). Asian dust was generated by the strong northwesterlies in the northern part of China at 09 LST on 10 November, and transported to the Korean Peninsula after 36 hours. The prediction was confirmed by the ground dust measurements over the Korean Peninsula in Fig. 11.3.3. Several PM_{10} stations recorded 1200~1800 μg/m^3 with the peak values moving from the west to the east: around 19 LST on 11 November at Baengnyeongdo, midnight at Seoul, and 5 LST on the following morning at Chupungnyeong.

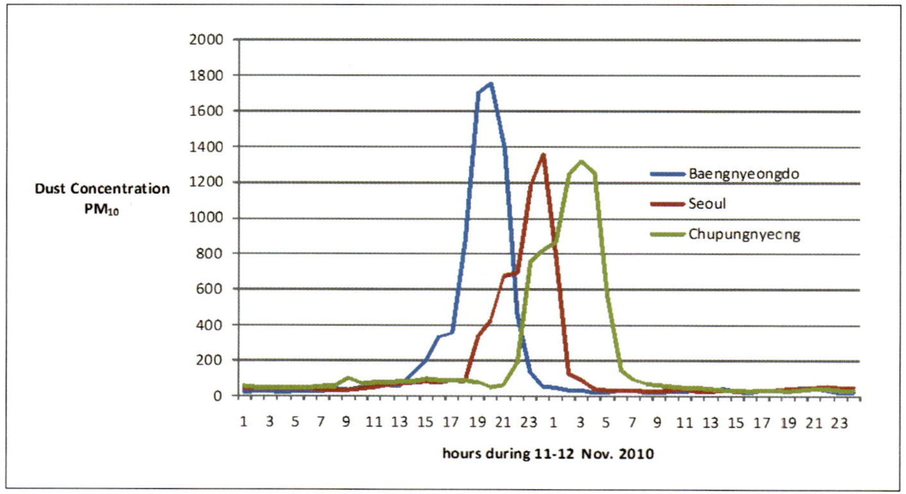

Fig. 11.3.3. Observed PM_{10} in units of μg/m^3 during 11~12 November 2010 at Baengnyeongdo Island, Kwanaksan Mountain, and Chupungnyeong in Korea. Source: KMA.

Users have to carefully monitor atmospheric conditions conducive to dust generation, and confirm if the observed dust concentration from ground and satellite measurements is consistent with the model forecast. The monitoring of surface conditions over the source region helps identify any discrepancy between the model forecasts and reality.

Wind varies greatly with height, and the forecast trajectory is sensitive to the starting height of maximum dust concentration in the source region. An ensemble forecast with various starting trajectories can demonstrate possible pathways leading to transboundary air pollution. Stronger wind and wavier trajectories result in larger forecast errors.

Dust interact with cloud particles and rain droplets, and models are poor at describing the physical processes of dry and wet deposition. Recall that clouds are the most unpredictable elements in a model. The errors in the model clouds lead to large errors in the dust concentration forecast along the dust trajectory.

12. Temperature, Wind, and Ocean Waves

12.1 Sensible weather near the surface

Sensible weather elements are greatly affected by the turbulent exchange of heat and momentum in the PBL bounded at about 1,500 m from the ground. The lower part of the PBL is mediated by the constant-flux surface layer immediately above the ground. Surface temperature depends on surface characteristics, which determine surface reflectivity or albedo of solar radiation, heat capacity and thermal conductivity of the soil. As the spatial grid resolution is not sufficient to resolve local surface characteristics, model forecast for surface parameters often deviates from the observation. The model surface albedo is obtained from the areal average over neighboring grid points, and may not be representative of either the water or land. In such a case, the diurnal range of the model surface temperature tends to be smaller than the observed over a land site and higher than the observed over a water body. An extreme example is presented in the lower panel of Fig. 12.1.1 for a grid point over the western island surrounded by the Yellow Sea. The model diurnal range over the island is underestimated by more than 3 degrees compared with the range corrected by statistical methods.

Latent and sensible heat fluxes are highly influenced by soil moisture holding capacity and rainfall amount. As soil water content increases after rainfall events, the surface loses more energy in the form of latent heat than sensible heat. The diurnal range of the surface air temperature decreases as a result. Photosynthesis of vegetation during the daytime allows latent heat release. Sparse vegetation at model grid points results in overestimation of the sensible heat flux and accordingly, of the maximum surface air temperature. Excessive soil moisture at the model grid points leads to increase in the latent heat flux and overproduction of clouds and precipitation. Advanced modeling centers periodically update the ancillary dataset including soil types with the help of satellite measurements. As the vegetation field changes with seasons, one should take into account the corresponding changes in the surface albedo and soil moisture.

Snow has a higher albedo than bare ground. Part of the incoming solar radiation is used to melt the snow instead of warming the ground. In addition, water vapor is released into the atmosphere during sublimation. After a recent snowfall, the daytime maximum temperature and the boundary layer moisture content in the model need to be adjusted towards lower and higher values respectively, unless the snow cover distribution is updated at the lower model boundary.

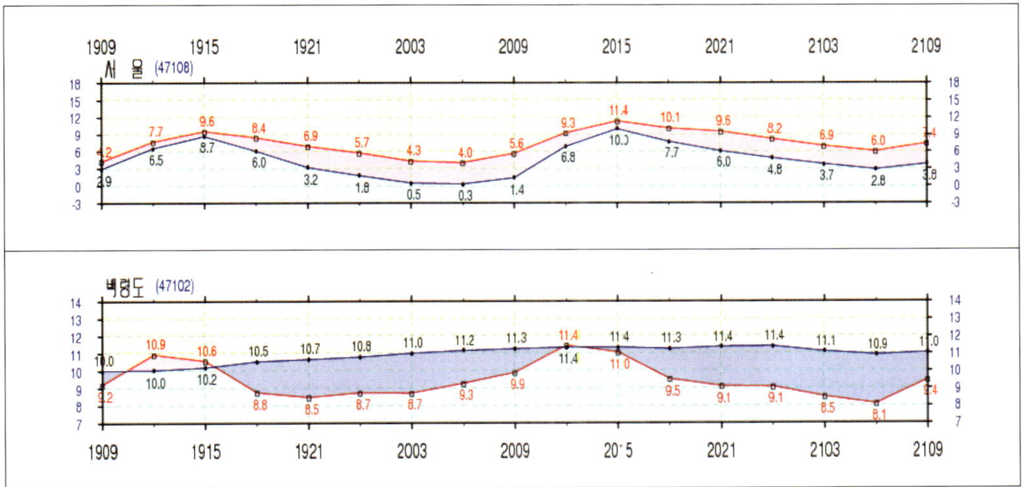

Fig. 12.1.1. Typical diurnal cycle for a typical land surface (upper) and island or near a water body (lower). The blue curve represents the direct model output derived from the UM model with a 12-km grid spacing. The red curve indicates the Kalman Filter bias-calibrated. Source: KMA.

Cloud-radiation feedback

Understanding the underlying principles of model physics and the associated model bias helps one interpret model forecasts of temperature and wind. The radiative transfer process in a model is highly sophisticated and little can be added by human forecasters. The model, however, is limited in predicting cloud properties and their interaction with other physical elements, and a meaningful correction for the model bias can be made in light of cloud-radiation feedback.

Clouds have dual role. Particularly at lower levels, they reflect incoming solar radiation back to space, resulting in reduced surface heating during the daytime. At the same time, they absorb terrestrial radiation and emit part of their radiation back to the ground. Higher-level cirrus clouds are more effective at shielding longwave radiation, keeping it from escaping out of the atmosphere at night.

If a model tends to overestimate the amounts of cloud cover and moisture, the surface maximum temperature needs to be adjusted toward higher values during the daytime, and lower values at night. Adjustments should be made in the opposite direction in the case of underestimation.

The PBL interacts with clouds, and modifies the vertical profiles of temperature, moisture, and wind. Shallow convective clouds capped by an inversion that developed through turbulent mixing of moist air in a stable boundary layer facilitate the ventilation of moisture out of the PBL. Underestimation of shallow convective clouds in a model results in overprediction of the amount of moisture in the PBL and the associated precipitation.

12.2 Wind and ocean waves

Surface wind depends on PBL height and surface roughness. A deeper PBL leads to downward transport of stronger wind and cooler air from the free atmosphere. If a model underestimates the PBL top, the model surface wind must be corrected to higher values. On the other hand, surface roughness exerts frictional drag on the wind force. The strength of a sea breeze decreases over land as the roughness length increases. Over rugged terrain, the surface wind is further modulated by the local topography. The model surface wind should be critically reviewed to see if any deficiency arises from poor representation of topography in the model surface. This point will be discussed further in Chapter 13.

An ocean wave model is forced with the lowest-level wind stresses produced by the atmospheric model. Thus the performance of a wave model depends on the accuracy of the wind forecasts. It is noted that wind observation is poor over the ocean, and heavily relies on satellite measurements. If the wind forecast deviates from reality, a systematic correction has to be made on the wave forecast accordingly. Swells, i.e., waves generated from distant storms with a significant wave height, travel out of the wave source region. The model track and intensity of the wave source are important error sources for swell height forecasts. Figure 12.2.1 presents the Wave Watch 3 (WW3) model forecasts for the wave height, frequency, and propagation vector. The high waves over the south of Japan and the south China Sea consist of long waves with periods over 9 seconds. They propagate to the southeast with the strong northerlies.

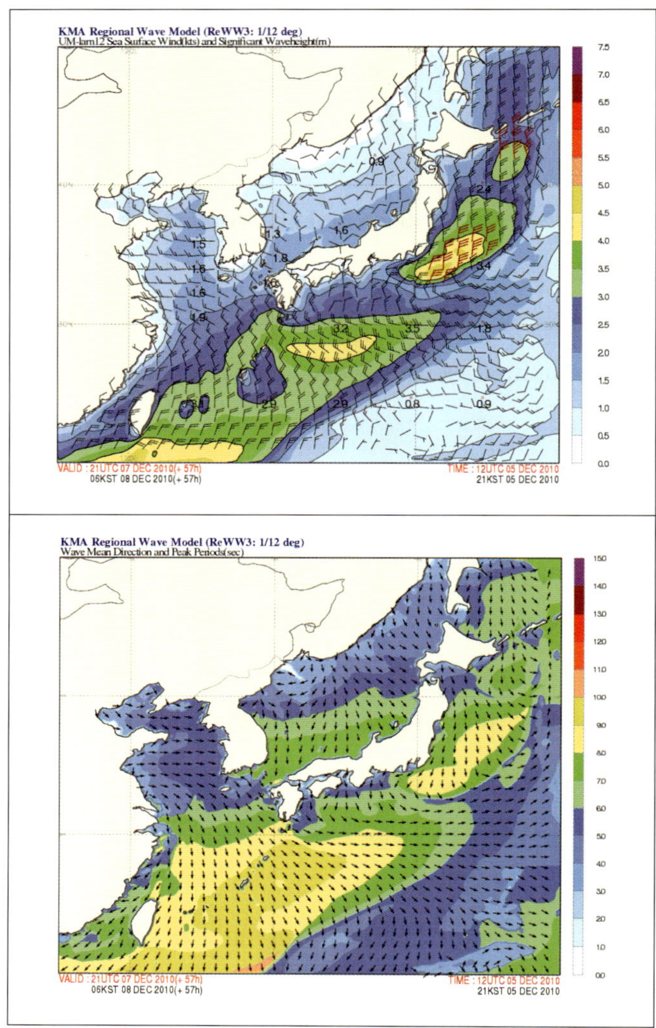

Fig. 12.2.1. A 57-hr forecast from the WW3 wave model for wave height (color) in units of m (upper), mean wave propagation vector in units of kt and peak periods (color) in units of seconds (lower). The model is forced by the surface wind stresses from the UM model with the initial state at 21 LST on 5 December 2010. Source: KMA.

13. Local Effects

The previous chapters already discussed the limitations of computer models in the prediction of local thunderstorms and mesoscale complex systems. There are other local phenomena that can hardly be identified in computer-generated prognostic charts. Even though the latest high-resolution mesoscale models may capture part of the local phenomena, the model forecasts often depart from reality in terms of timing, intensity and locality. A re-examination of model outputs by human forecasters is essential to correct their bias and properly forecast local weather. The COMET modules developed at NCAR provide practical guides with a handful of conceptual models for various mesoscale phenomena. Two commonly observed local features are reviewed in this chapter to help readers appreciate the complexity of fine-scale weather and the associated forecasting challenges: atmospheric phenomena over mountain and coastal zones.

13.1 Mountain zones

The weather and climate on mountains are quite different from what we observe in the plains. The higher the elevation, the higher the risk of sunburn, and the colder, windier, and wetter the weather. Precipitation and fog are enhanced by mechanically-induced upward motions on the windward slope. In contrast, dry conditions prevail on the downwind side of a mountain.

Incident ultraviolet radiation increases about 6% per kilometer of altitude in clear sky conditions according to the U.S. EPA. Normally, air temperature cools about 6.5 degrees for every 1 km of ascent. The local geography and vertical stability affect the cooling rate, as demonstrated by the measurements at the central Sierra Nevada in Fig. 13.1.1. The increase of wind speed with height in general fits with the logarithmic law in the lowest 100 m, i.e., roughly a 3.6 km/hr increase for every ascent of 10 m in the neutral stability case. The details of wind profiles depend on the thermal stability of the lower atmosphere. Near complex terrains, valley-mountain circulation further complicates the logarithmic wind relationship. At night the mountain breeze sets up. Cold air over the mountain top is drawn downward, causing much colder temperatures on winter nights.

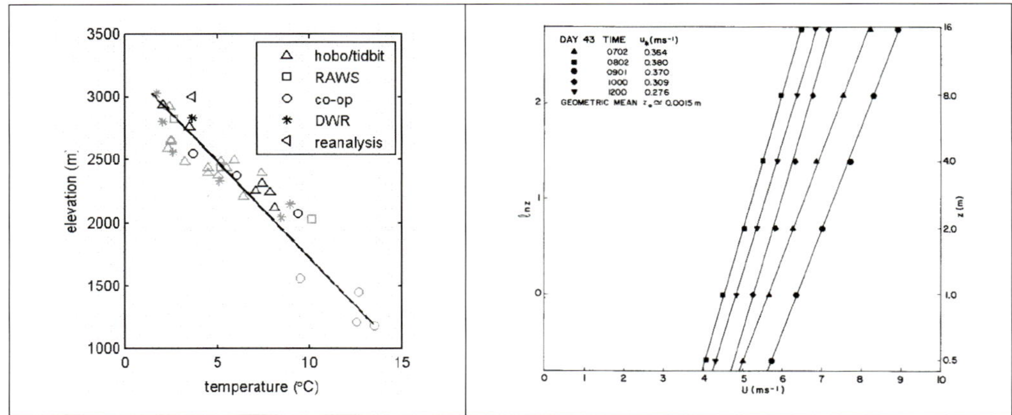

Fig. 13.1.1. (Left) Average temperature over mountains at each measured location between 17 July 2002 and 30 June 2004, compared with the standard lapse rate of -6.5°C per km. The gray symbols represent west slope sites; the black symbols represent east slope sites (Lundquist and Cayan, 2007). The different stations or sensor types are shown in the legend. (Right) Comparison of the observed wind profiles in the neutral surface layer of the Wangara experiment at varying hours and roughness lengths. The wind increases by roughly 3.6 km/hr for every 10 m increase in height in the neutral case. This figure is taken from http://www.heliosat3.de/e-learning/BLM/lect5.pdf.

For mountain ranges elongated in the north-south direction, cold air approaching from the east accumulates over the upslope areas and a stable high-pressure system builds up. The pressure gradient force gradually pushes the cold air mass down to the low latitudes, damming cold air. The eastern slopes of major mountain ranges are favorable to cold air damming: east of the Rockies, the eastern coast of the United States, east of the Andes in Argentina, the eastern coast of Brazil, east of the Alps, the eastern coast of Greece, eastern coasts of Korea and Japan, and the eastern coast of the Kamchatka Peninsula. Cold air damming often results in extensive cloud cover and dense fog for an extended period. It also provides conditions for sleet and freezing rain that cause severe damage on crops, material possessions, trees, and power lines (Bailey et al., 2003). Figure 13.1.2 shows an example over the eastern mountain in the Korean Peninsula. Wind direction readings across the mountain ridge help correct model forecasts of precipitation, wind strength, and temperature.

Fig. 13.1.2. Cold air damming frequently occurs on the eastern slopes of mountain ranges in high latitudes where polar air masses migrate equatorward by the force of gravity. The cloud image is taken from the visible channel of MTSAT-1R over the Taebaek mountain range in Korea at 06 UTC on 23 October 2006. Source: KMA.

When a cross-barrier flow passes over a mountainous terrain with a Froude number near 1 and an inversion just above the mountain top, a downslope wind develops with peak gusts that may easily exceed twice the sustained wind speed. As the wind speed over the mountain top increases, the downslope wind increases as well. Records show that downslope wind speeds may reach 160 km/hr. The cold downslope wind is called Bora at the Adriatic coast, Mistral at the French

Riviera, and Fall Winds in Greenland and Antarctica. It often starts in snow-covered plateaus where the air is chilled when it comes in contact with a cold surface, and is pulled downslope by gravity.

Warm downslope wind occurs as humid air that loses moisture by precipitation on the upslope is brought downslope by adiabatic warming. Figure 13.1.3 is a schematic of a Foehn wind. The temperature in Rapid City USA rose from -20°C to 7°C in two minutes. This type of wind is called Chinook over the Rockies, Foehn over the Alps, Puelche on the west slopes of the Andes, Zonda on the east slope of the Andes, and Yanganjipung on the east slope of the Taebaek mountains in Korea. Downslope winds are also found in other parts of the world including the Koroglu Mountains in Turkey, the Tibesti Mountains in Africa, the Barisan Mountains in Indonesia, and the Southern Alps of New Zealand.

Mesoscale models can capture gross features of the downslope wind and associated temperature changes downstream. Wind direction readings with respect to the ridge, strength and height of inversion, and knowledge of local geography help compensate for the limitations of model predictions. More details on cold-air damming and downslope winds are available from the COMET modules.

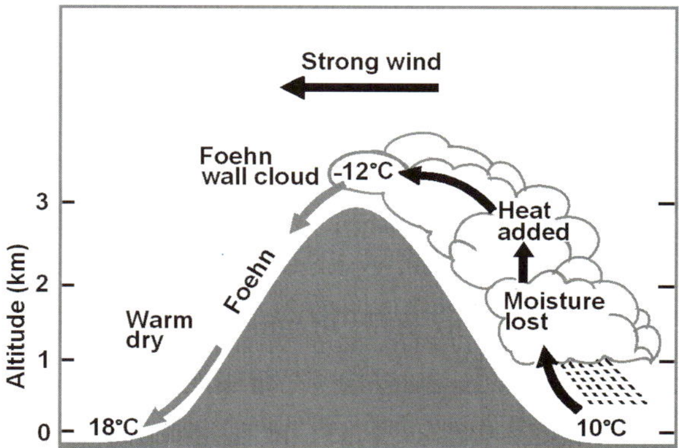

Fig. 13.1.3. Schematic diagram for the Chinook or Foehn downslope winds that are observed in many parts of the world. Each region has different names for the downslope wind. The figure is reproduced from the COMET module (http://www.unca.edu/~dmiller/Chapter%207.ppt)

A local wind blowing through a gap between mountains is called a gap wind. Drag is increased as the air enters a narrow channel. Its speed is slowed down by a blocking effect near the entrance. Air accumulates, and pressure increases. Conversely, air spreads out near the exit, and pressure decreases. The resultant pressure gradient force supports the acceleration of wind speed downstream near the exit. In addition, air passing through the narrow channel gains speed by the Bernoulli effect. Figure 13.1.4 presents an example of a gap wind. The enhanced northwesterlies in red indicate the gap winds that pass through the valleys. Famous regions for gap winds include the Strait of Juan de Fuca, Canada, Tehuantepecer near the Panama Canal, the Strait of Gibraltar, Squamish in British Columbia, and Hinlopenstretet near the Arctic. A gap wind can occur downstream of a cape, with low-level inversion acting as a wave guide for the wind blowing parallel to the coastal barriers. Small-scale gap winds are found anywhere with a rugged terrain.

Very-high-resolution mesoscale models can capture the main features of a gap wind. The current operational models with a grid spacing of 5~10 km apart are too crude to resolve a gap wind over a complex terrain. Readings of wind direction, stability, and local knowledge of geography are essential to correct the model forecasts.

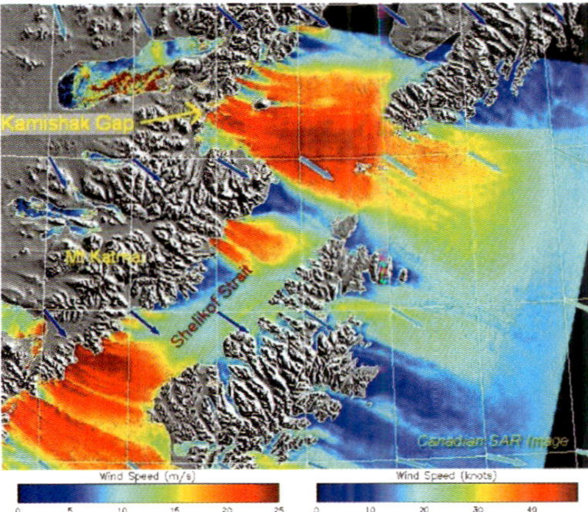

Fig. 13.1.4. A strong synoptic northwest wind event near the Lower Cook Inlet and Kodiak Waters at 17 UTC on 12 December 1999, from the SAR (space-borne synthetic aperture radar)-based surface

wind speed fields, with a 200-m grid spacing (Zingone and Hufford, 2006). The wind speed scales are presented in both m/s and kts. The image is taken from http://pafc.arh.noaa.gov/papers/Analyzing_Surface_Wind-AMS.pdf. Source: NASA.

13.2 Coastal zones

Coastal zones reveal unique climatology, as they are under the influence of adjacent water bodies. Sea breezes occur and penetrate inland during the daytime when the land surface warms up more rapidly than the sea surface. Conversely, land breezes form at night when the sea surface preserves moderate heat while the land cools rapidly by upward infrared radiation. Wind surfers enjoy sea breezes on a sunny day as they blow constantly toward the coast. A land sea breeze is most evident when large-scale forcing is weak. Squall lines often develop along a sea breeze front propagating inland for a few tens of kilometers after picking up abundant moisture from the ocean. The sea breeze in Florida State in Fig. 13.2.1 is just one example of what can occur in any part of the world. Wind direction readings across the shore from the weather map are critical for the forecast of local temperature and thunderstorm development.

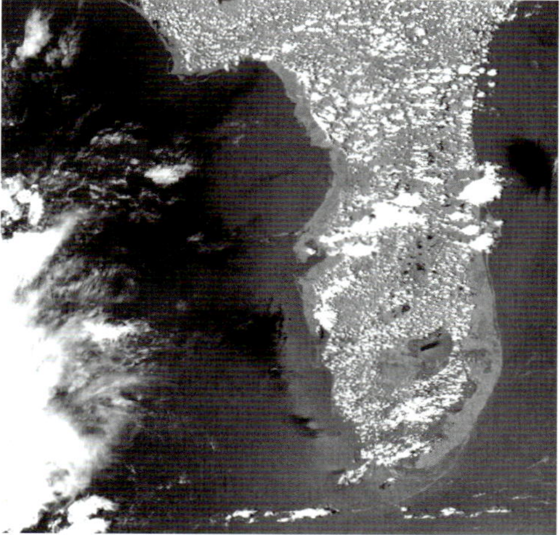

Fig. 13.2.1. An example of a sea breeze in Florida State, U.S.A. as shown in the Advanced Very High

Resolution Radiometer, AVHRR image. Clouds form along with onshore flow around the coast. The image is available from http://itg1.meteor.wisc.edu/wxwise/satmet/lesson7/seabreezetrack.html.

Cold water bodies in the western region of major continents are maintained by the upwelling of deep water, as induced by offshore ocean currents at the flank of a permanent high. These regions are favorable to the development of fog and stratus clouds as warm and moist air can easily condense above the cool ocean under stable conditions. Fog also develops frequently over the Northwest Atlantic, Northwest Pacific, and Southwest Atlantic during summer when southwesterly winds flow over the relatively cool ocean surface along the western rim of a permanent high. Readings of onshore winds over the relatively cool surface from the weather map help forecast the presence of fog or stratus clouds.

Sometimes, the land-sea contrast sets up a strong thermal gradient off the coast, and is often enhanced by high mountains with daylight heating and surrounding cold water. Strong low-level jets, called coastal jets, develop along the coastline in parallel with the isotherms. Synoptic condition favorable to coastal jets can be found in late summer over the west coasts of California, South America, South Africa, and the southern coast of Saudi Arabia. More details on coastal fog and coastal jets are available in the COMET modules.

Coastal zones are also vulnerable to storm surges caused by tropical cyclones, and more importantly by tsunamis or seismic sea waves. Tsunamis are long but shallow waves generated by an abrupt rise in sea level during an earthquake below the bottom of the sea. The waves, however, can easily turn into higher than 10-story towers of water as they approach the coast during an earthquake with moderate magnitude. As tsunamis travel very fast, i.e., at a speed of 800 km per hour, their abrupt presence threatens coastal communities. The 2004 Indonesian tsunami resulted in 230,000 casualties, and about $4 billion US dollars in insured losses. Areas at great risk are those less than 15m above sea level and within 1.5~5 km of the shoreline. If one hears an official tsunami warning or detects signs of a tsunami, he or she should evacuate at once and climb onto higher ground. Tsunamis cannot be seen on a weather map.

For the latest tsunami warnings and more information, visit the website of the International Tsunami Information Centre of the Intergovernmental Oceanographic Commission, the Pacific Tsunami Warning Centre based in Hawaii, or the West Coast and Alaska Tsunami Warning Centre based in Alaska:

http://ioc3.unesco.org/itic/,

http://www.prh.noaa.gov/ptwc/, and

http://wcatwc.arh.noaa.gov/.

References

Bailey, C. M., G. Hartfield, G. M. Lackmann, K. Keeter, and S. Sharp, 2003: An objective climatology, classification scheme, and assessment of sensible weather impacts for Appalachian cold-air damming. *Wea. Forecasting*, **18**, 641-661.

Baxter, M. A., C. E. Graves, and J. T. Moore, 2005: A climatology of snow-to-liquid ratio for the contiguous United States. *Wea. Forecasting*, **20**, 729-744.

Bourgouin, P., 2000: A method to determine precipitation types. *Wea. Forecasting*, **15**, 583-592.

Chernykh, I. V., and R. E. Eskridge, 1996: Determine of cloud amount and level from radiosonde soundings. *J. Appl. Meteor.,* **35**, 1362-1369.

Choi, J.-T., Y.-K. Seo, and G.-B. Jin, 2005: Determining algorithm of precipitation types and sky conditions for digital weather forecast. *J. Meteorl. Tech.*, **1**, 25-31.

Coiffier, J., 2004: Forecasting standards- weather forecasting technique considered as a sequence of standard processes from the forecaster's point of view. Working document of CBS-ICT/DPFS (WMO).

COMET, 2010: *Effective use of high-resolution models.* University Cooperation for Atmospheric Research.

Doswell, C. A., III, 1986: *Mesoscale meteorology and forecasting*, American Meteorological Society, Boston, pp. 689 - 719.

Elsberry, R. L., and L. E. Carr III, 2000: Consensus of dynamical tropical cyclone track forecasts - errors versus spread. *Mon. Wea. Rev.*, **128**, 4131-4138.

Fritsch, J. M., and R. E. Carbone, 2004: Improving quantitative precipitation forecasts in the warm season: a USWRP research and development strategy. *Bull. Amer. Meteorl. Soc.*, **85**, 955-965.

Froude, 2009: Regional differences in the prediction of extratropical cyclones by the ECMWF ensemble prediction system. *Mon. Wea. Rev.*, **137**, 893-911.

Gilleland, E., D. A. Ahijevych, B. G. Brown, and E. E. Ebert: 2010: Verifying forecasts spatially. *Bull. Amer. Meteorl. Soc.*, **Oct.**, 1365-1373.

Goerss, J. S., 2007: Prediction of consensus tropical cyclone track forecast error. *Mon. Wea. Rev.*, **135**, 1985-1993.

Guard, C. P., and M. A. Lander, 1999: *A scale relating tropical cyclone wind speed to potential damage for the tropical Pacific Ocean region: A user's manual.* WERI Technical Report 86, University of Guam, Mangilao, Guam, 60 pp.

Harper, B. A., J. D. Kepert, and J. D. Ginger, 2008: *Guidelines for converting between various wind averaging periods in tropical cyclone conditions*, WMO, 51 pp.

Harper. K., L. W. Uccellini, E. Kalnay, K. Carey, and L. Morone, 2007: 50th anniversary of operational numerical weather prediction. *Bull. Amer. Meteorl. Soc.*, May, 639-650.

Harr, P., 2010: *Formation forecasting.* 7th International Workshop on Tropical Cyclone, 15-20 November 2010, La Réunion.

Herndon, D., C. S. Velden, J. Hawkins, T. Olander, and A. Wimmers, 2010: *The CIMSS SATellite CONsensus (SATCON) tropical cyclone intensity algorithm.* 29th Conf. Hurricanes and Tropical Meteor., Amer. Meteor. Soc., Tucson, AZ, 10-14 May.

Hoskins, B. J., M. E. McIntyre, and A. W. Robertson, 1985: On the use and significance of isentropic potential vorticity maps. *Quart. J. Roy. Meteor. Soc.*, **111**, 877-946.

Hsieh, W. W., A. Wu, and A. Shabbar, 2006: Nonlinear atmospheric teleconnections. *Geophysical Research Letters*, **33**, L07714, doi:10.1029/2005GL025471.

IPCC, 2007: *Climate change : Synthesys report, summary for policy makers.* 22pp.

Kang, H., K. An, C. Park, A. L. S. Solis, and K. Stitthichivapak, 2007: Multimodel output statistical downscaling prediction of precipitation in the Philippines and Thailand, *Geophys. Res. Lett.*, **34**, L15710, doi:10.1029/2007GL 030730

Kleinknecht, A., 2002: Afraid of the weather? *Weatherwise.* Nov/Dec. 14-20

Kocin, P. J., and I. W. Uccellini, 2004: A snowfall impact scale derived from northeast storm snowfall distributions. *Bull. Amer. Meteorl. Soc.*, Feb. 177-194.

Landsberg, H. E., 1986: Weather, climate and you. *Weatherwise.* October, 248-253

Leigh, R., 2007: *Hail storm- One of the costliest natural hazards.* Coastal cities natural disasters, 20-21 Feb. 2007, 1-8.

Lundquist, J. D., and D. R. Cayan (2007), Surface temperature patterns in complex terrain: Daily variations and long-term change in the central Sierra Nevada, California, *J. Geophys. Res.*, **112**, D11124, doi:10.1029/2006 JD 007561.

Majumdar, S. J., and P. M. Finocchio, 2010: On the ability of global ensemble prediction systems to predict tropical cyclone track probabilities. *Wea. Forecasting*, **25**, 659-680.

Martin, J. D., and W. M. Gray, 1993: Tropical cyclone observation and forecasting with and without reconnaissance. *Wea. Forecasting*, **8**, 519-531.

Matsuo, T., and Y. Sasyo, 1980: Non-melting phenomena of snowflakes observed in sub saturated air below freezing level. *J. Meteorl. Soc. Japan*, **59**, 26-32.

Meehl, G. A., 1987: The annual cycle and interannual variability in the tropical Pacific and Indian Ocean regions. *Mon. Wea. Rev.*, **115**, 27-50.

Moncrieff, M. W., M. A. Shapiro, J. M. Slingo, and F. Molteni, 2007: Collaborative research at the intersection of weather and climate. *WMO Bulletin*, **56**, 204-211

Morinari, J., and E. Fukada, 2010: *Observation capabilities and opportunities.* 7th international workshop on tropical cyclone, La Reunion, 15-20 Nov.

Mullen, S. L., and R. Buizza, 2001: Quantitative precipitation forecasts over the United States by the ECMWF ensemble prediction system. *Mon. Wea. Rev.*, **129**, 638-663.

National Research Council, 2010: *When weather matters: science and service to meet critical societal needs.* National Academies Press, 208 pp.

Nguyen, V-T-V., T-D. Nguyen, and F. Ashkar, 2002: Regional frequency analysis of extreme rainfalls. *Water Science and Technology*, **45**, 75-81.

Ohfuchi, W., H. Sasaki, Y. Masumoto, and H. Nakamura, 2007: "VIRTUAL" atmospheric and oceanic circulation in the earth simulator. *Bull. Amer. Meteorl. Soc.*, June, 861-866.

Osczevski, R. and M. Bluestein, 2005: The new wind chill equivalent temperature chart. *Bull. Amer. Meteorl. Soc.*, Oct. 1453-1458

Palmer, T. N., 2003: *Predicting uncertainty in forecasts of weather and climate.* ECMWF Meteorological Training Course Lecture Series, 48pp.

Park, S.-U., A. Choe, M.-S. Park, and Y. Chun, 2010: Performance Tests of the Asian Dust Aerosol Model 2 (ADAM 2). *Journal of Sustainable Energy & Environment*, **1**, 77-83

Persson, A., 2001: *User Guide to ECMWF forecast products.* Meteorological Bulletin, M3.2, ECMWF, 123pp.

Ramage, C. S., 1995: *Forecasters guide to tropical meteorology.* AWS/TR-95/001, Air Weather Service, 392 pp.

Riddaway, R.W., 2002: *Numerical methods.* ECMWF meteorological training course lecture series. 98pp.

Stern, H., 2005: *Future role of the human in the forecast process – to provide input to a system that mechanically integrates judgmental (human) and automated predictions?* 21st Conference on Weather Analysis and Forecasting; 17th Conference on Numerical Weather Prediction; Amer. Meteor. Soc., Washington, DC, 1-5 August, 2005.

Sznaider, R. J., 2005: *Operational uses of weather information in GIS-based decision support systems*, 19 pp.

Vanicek, K., T. Frei, Z. Litynska, and A. Schmalwieser, 1999: *UV-index for the public.* COST-713 Action, Brussels.

Webster, P. J., 1983: *The large scale structure of the tropical atmosphere.* In Hoskins and Pearce (eds.) General Circulation of the Atmosphere, London: Academic Press. 235 – 275pp.

WMO, 1993: *Guide on the Global Data-processing System*, WMO N° 305

Zingone, E., and G. L. Hufford, 2006: *Analyzing surface wind fields near lower Cook Inlet and Kodiak Waters using SAR.* Alaska Marine Science Symposium, J4.11

Subjective Index

Only the most important discussions of each subject are indexed.

Woo-Jin Lee

Woo-Jin Lee is Director-General of the Numerical Prediction Office at Korea Meteorological Administration. He received a Ph. D in atmospheric sciences from the University of Illinois at Urbana-Champaign. Between 1998 and 2010, he held a series of senior posts in the Korea Meteorological Administration, with particular focus on numerical weather prediction and weather forecasting. He participated on various programs under the World Meteorological Organization and Typhoon Committee, and served as Director of APEC Climate Center. He has published four books, Meteorological Services for Information Society (1997, in Korean), Interpretation of Weather Charts (2007, in Korean), Weather Prediction with Supercomputers (2007, in Korean), Weather of Korea - A synoptic Climatology (2010), and numerous articles in meteorological journals.